Charles C. Miller

A Year among the Bees

Being a Talk about some of the Implements, Plans and Practices of a

Bee-Keeper of 25 Years' Experience

Charles C. Miller

A Year among the Bees
Being a Talk about some of the Implements, Plans and Practices of a Bee-Keeper of 25 Years' Experience

ISBN/EAN: 9783337337247

Printed in Europe, USA, Canada, Australia, Japan

Cover: Foto ©berggeist007 / pixelio.de

More available books at **www.hansebooks.com**

Among the Bees;

BEING

A Talk about some of the Implements, Plans and Practices of a Bee-keeper of 25 years' experience, who has for 8 years made the Production of Honey his Exclusive Business.

By C. C. MILLER, M.D.

CHICAGO, ILL.:

OFFICE OF THE AMERICAN BEE JOURNAL,

923 and 925 West Madison Street.

A Year among the Bees.

INTRODUCTION.

One morning, five or six of us, who had occupied the same bed-room the previous night during the North American Convention at Cincinnati, in 1882, were dressing preparatory to another day's work. Among the rest were Bingham, of smoker fame, and Vandervort, the foundation-mill man. I think it was Prof. Cook who was chaffing these inventors, saying something to the effect that they were always at work studying how to get up something different from anybody else, and, if they needed an implement, would spend a dollar and a day's time to get up one "of their own make," rather than pay 25 cents for a better one ready-made. Vandervort, who sat contemplatively rubbing his shins, dryly replied : "But they take a world of comfort in it." I think all bee-keepers are possessed of more or less of the same spirit. Their own inventions and plans seem best to them, and in many cases they are right, to the extent that two of them, having almost opposite plans, would both be losers to exchange plans.

In visiting and talking with other bee-keepers I am generally prejudiced enough to think my plans are, on the whole, better than theirs, and yet I am always very much interested to know just how they manage, especially as to the little details of common operations, and occasionally I find something so manifestly better that my own way, that I am compelled to throw aside my prejudice and adopt their better way. I suppose there are a good many like myself, so I think there may be those who will be interested in these bee-talks, wherein I shall try to tell honestly just how I do, talking in a familiar manner, without feeling obliged to say " we " when I mean " I." Indeed I shall claim tho privilege of putting in the pronoun of the first person as often as I please, and if the printer runs out of big I's toward the last of the book, he can put in little i's.

Moreover, I don't mean to undertake to lay down a methodical system of bee-keeping, whereby one with no knowledge of the business can learn in " twelve short lessons " all about it, but will just talk about some of the things that I think would interest you, if we were sitting down together for a familiar chat. I take it you are familiar with the good books and periodicals that we as bee-keepers are blessed with, and in some things, if not most, you are a better bee-keeper than I; so you have my full permission, as you go from page to page, to make such remarks as, " Oh, how foolish ! " " I know a good deal better way than that," etc., but I hope some may find a hint here and there that may prove useful.

I have no expectation nor desire to write a complete treatise on bee-keeping. Many important matters connected with the art I do not mention at all, because they have not come within my own experience. Others that have come within my experience I do not mention, because I suppose the reader to be already familiar with them. I merely try to talk about such things as I think a brother bee-keeper would be most interested in if he should remain with me during the year.

As I want to get down to bee-talk as quickly as possible, I think I'll let this serve for both preface and introduction. As for dedication, there are lots of good friends I'd like to dedicate my little work to, but I hardly like to single out any one of them, unless it should be my blessed old mother, and she hardly knows a drone from a queen, so that would hardly do. On the whole, I think I'll not dedicate it to any one.

TAKING BEES OUT OF THE CELLAR.

The difficulty of wintering bees, at the north, is not entirely without its compensations. I am almost willing to meet some losses, for the sake of the sharp interest with which I look forward to the time of taking the bees out of the cellar in the spring. I live on a place of 37 acres, about a mile from the railroad station, and on my way down town there is a soft-maple tree which blooms in advance of all others, at least a day or two. How eagerly I watch that tree from the first bursting of the buds, and when the red of the blossom actually begins to push forth, with what a thrill of pleasure I say, " The bees can get out on the first good day !"

In former years I did sometimes bring out the bees earlier, because they seemed so uneasy, but I doubt if I gained anything by it. I have known one or two years when a cold, freezing time came on at the time of maple bloom and I did not take out the bees for a good many days, but generally I go by the blooming of the soft maples. So I watch the thermometer and the clouds, and usually in a day or two there comes a morning, with the sun shining, and the mercury at 45° or 50°. This is one of the few times when I call in outside help ; for I want to make sure of getting out all the bees I can on the first warm day. So I leave word with a neighbor, Moses Dimon, or Mr. Tyler, who lives in my tenant house, the evening before, to be on hand in the morning if the weather is fine. My only son, Charlie, nearly 18 years old, is the best help I ever had at carrying bees, indeed, at a good deal of the bee-work, but he doesn't take kindly to the business. When little, he did not care much for a bee-sting, but a few years ago he got so that a sting made him spotted all over, even his tongue and ears swelling up, causing great suffering. I think he has entirely outgrown this difficulty, but I am afraid he never will overcome the dread of a sting.

Some object to carrying out many colonies at a time, for fear of their swarming-out, from the excitement of so many flying at once, but I have never had any difficulty in this direction. The evening before, I open all cellar doors and windows, and although the bees may roar for some time, usually they will be very quiet next morning. I generally stay in the cellar and help lift the hives off the pile. The entrances are left open just as they were all winter, the hives are carried quietly and set upon the farthest stands first, to avoid passing by them afterward. The covers are at once put on, and generally it will be some little time before they start to flying. Their remaining so quiet is partially the result of their being handled carefully, but is mainly due, I think, to the thorough airing they have had in the cellar.

Before each hive leaves the cellar, I make sure there are live bees in it by placing my ear to it, or lifting one end of the quilt. If any are dead, they are piled up to one side in the cellar. The stands of the home apiary are all filled, and the remaining hives are piled three high on a couple of sticks of firewood, not far from the cellar door. This is done so as to occupy little room, and necessitate their being carried a shorter distance when being put on the wagon to be hauled to the out apiary. No attention whatever is paid to numbers of hives on carrying out. Entrances are nearly closed after the first flight.

NUMBERING HIVES.

Numbers for hives are made in this way : Pieces of tin 4 by 2½ inches have a small hole punched in each one, near the edge, about midway of one of the longer sides. With ½-inch wire nails, nail them on the top of a wooden hive-cover or other plane surface. Then give them a couple of coats of white paint, and, when dry, put the numbers on them, from 1 upward, with black paint. There is room to make figures large enough to be seen distinctly at quite a distance. I fasten these tin-tags on the front of the hive, by *pushing* a ½ or ⅝-inch wire nail through the hole into the hive, with a cold-chisel. This does not disturb the bees as driving with a hammer would. I can easily take them off the hive by slipping the cold-chisel under the tin and prying up.

When the hives are put on the stands in the spring, the numbers are all mixed up. The first thing to be done is to enter upon the record-book these numbers. The first hive in the back row should be No. 1, the next No. 2, and so on; but in the place of No. 1 stands perhaps 231, on the place of No. 2 stands 174, etc. So, on the new record-book I write No. 1 (231) on the first page at the top; one-third the way down the page, I write No 2 (174), and so on.

HAULING BEES.

As soon as the bees have had a good flight, those that do not belong to the home apiary are ready to be hauled away. I like to get them away as soon as possible, but sometimes the state of the roads makes delay advisable. I haul them on a one-horse wagon, putting 3 hives into the wagon-box; then a light rack, something like a hay-rack, on the box, holds eight more, making eleven at a load. A two-horse wagon with a hay-rack holds about twice as many, and I sometimes use one in hauling, but as I keep only one horse I generally use him alone. I formerly supposed that bees must be hauled on springs or on a couple of feet of hay, but Mr. T. L. Von Dorn, of Omaha, told me he hauled them on a hay-rack with neither springs nor hay, and since then I have done the same with perfect success. I don't like to disturb the frames of the hives in any way before hauling them, as they remain firmer in place if their fastenings of bee-glue remain undisturbed.

To prepare the hives for hauling, nothing is done to the inside of the hive, the cover is taken off, and a piece of cotton-cloth or sheeting, large enough to cover the hive and project two or three inches at each side and end, is put on the hive, and the cover replaced. Then the cover is tied on by means of a strong cord or sheep-twine, going under the front cleat, then crossing like the letter X on the top and going under the back cleat. The entrance is closed by means of wire-cloth in this way: I take a stick as long as the width of the hive or a little less, 1 inch in width, and ⅜ in thickness; on this I nail, with double-pointed tacks, a doubled piece of wire-cloth, so that the doubled edge shall project below the stick a half an inch. A single thickness of wire cloth would ravel. This projecting wire cloth covers the entrance to the hive, which is ½ inch high and the full width of the hive, and furnishes all the ventilation necessary, unless in very hot weather. The "stopper," as I call it, is fastened to the hive by driving a 1½-inch wire nail an inch or two from each end of the stick, and deep enough into the hive to hold firmly, then bending over the part of the nail not driven in, so that, as the nail-hole in the stick grows larger from frequent use, there will be no danger of the stick slipping back from its place.

Before putting the hives on the wagon, I examine each one carefully on all sides, above and below, to be sure that there is no possible chance for a bee to get out anywhere. I generally drive on a walk, but sometimes trot on a smooth piece of road. I have hammer and nails with me to provide against any contingency, and also a smoker, having the smoker lighted if I want to feel doubly secure. In case anything goes wrong on the road, which of late rarely happens, I unhitch the horse as quickly as possible, and leave him some distance from the wagon till everything is made secure. If a bee gets on the horse's head, the first impulse of the horse seems to be to get to some place where he can rub. I immediately spring to his head, and with hands and arms rub over his head and neck, taking care that, in his efforts to rub against me, he does not knock me over with his head.

When the hives are placed on their stands, in the out apiary, the two nails are drawn from the stoppers with a claw-hammer, and the stoppers very carefully removed. Sometimes I use a little smoke.

SPRING OVERHAULING.

Unless there is special reason for it, such as the fear of immediate starvation, no hive is opened until the bees have had one day for a good cleansing flight. This flight is usually taken on the day of setting out. Sometimes, however, a hive may be set out so late in the day that very few of its inmates fly till the next day, although I usually stop taking out, if I think a good flight will not be taken before night. After this cleansing flight is taken, the bees are ready to be overhauled on the first fine day, or the first day they can fly, for there may happen one or more days when bees cannot fly, and if frames of brood are taken out, the brood may be chilled. Besides, I am afraid it is not good for the bees themselves to be stirred up at such times, and it is not pleasant for the operator.

The bees that are taken to the out apiary on one day are generally overhauled the next. This out apiary is about 3 miles distant in a bee-line. It would, no doubt, be better farther away, but it is on the farm of John Wilson, my wife's father, who, with his "gude wife" Margaret, are rugged old Scotch people, and as my wife and her sister Emma are my principal assistants, it is so pleasant for them to make frequent visits to the spot where they were born, that I forego the advantage of having the apiary at a greater distance. I believe there might be an advantage in dividing up into a larger number of apiaries, and probably I shall act upon this belief.

But now let us proceed to the overhauling. I get my tool-box, bee hat and smoker, and go to hive No. 1, although the tag on it may say 231. Before this, however, an empty hive has been cleaned, more likely several; perhaps Charlie is cleaning them as fast as I use them.

After trying a number of different things for hive-cleaners, I have been best satisfied with a hatchet, the handle sawed short, so that it will not be in the way when working in the bottom of the hive, the edge very dull and a perfectly straight line, and the outside part of the blade also ground to a straight line and at right angles with the edge. This right-angled corner is to clean out the corners of the hive, especially the rabbets. In cleaning, the hatchet is moved rapidly back and forth, or rather from side to side, the blade being held at right angles to the surface being cleaned. The weight of the hatchet is quite a help, something like a fly-wheel in machinery. The propolis is scraped out of the hive and especial pains taken to clean the rabbets.

Having a hive ready, now for a seat. Bro. Doolittle once tried to poke a little fun at me in convention, because I accidentally admitted that I sat down to work at bees. If I were obliged to work all the season without a seat, I am afraid I would have to give up the business from exhaustion. Moreover, if I had the strength of a Samson I don't think I should waste it stooping over hives, so long as I could get a seat. I generally have three or four seats about the apiary, and I just take for a seat a box in which 500 sections have been shipped, whittling a hole in it to carry it by. By placing it differently, it gives me a seat of three different heights, suitable for working at a one-story hive, or one with supers tiered up on it.

Having placed my seat beside the hive to be overhauled, I put the empty hive beside it, the back end of the empty hive at the front end of the full hive, or else the front end of the empty hive at the back end of the full one. Lifting off the cover, I give one or two puffs of smoke at the entrance, then slowly peeling up the quilt with one hand, I blow a little smoke lightly, with the other, across the tops of the frames, not down into the cluster of bees. Still keeping the smoker in one hand, I pry up the frames at each end with a chisel made from an old file nearly a foot long; one end, the end I have just been using for prying loose the frames, being made square for about three inches of its length and brought to a blunt point, the other end like any cold-chisel, and about one inch wide. The frames will not need prying up again till toward fall, when the accumulation of propolis will make the chisel again necessary.

Mr. A. I. Root, I suppose, would say I ought to have metal corners to my frames and avoid the necessity of prying them loose. I tried two such hives made by him and I didn't like them. For one thing, my frames, as they are, are ready spring and fall, for hauling, without any nailing or fastening of any kind. Right here is a good place to enter a protest against the belief that I think my plans and fixtures are, under all circumstances, better than those of any one who differs from me. Mine may be best for me, and his for him. Or, it may be that if I knew enough about his, or had tried them *in the right way*, I would at once discard mine and adopt his. I think I am rather conservative about changes, old fogy if you like, but I expect to keep making changes so long as I keep bees.

Having finished using the chisel, or even if I should put it out of my hand for a minute, I must at once put it in the box. Formerly, I sometimes put a tool on the ground, and then I would forget where I put it, and perhaps not find it for two or three days. If, for any reason, I do not want to put it in the box, I put it on the top of a hive, so I can see it from a distance.

After loosening the frames, I lift out the one at the south side, (the hives face east), and place it in the empty hive ; then the next frame, and so on, watching for the first frame with brood. This frame of brood is placed next to the south frame in the new hive, but before putting it there I glance to see whether the queen is on it. I have always noticed that the bees, if left to themselves, have little or no brood in the south frame, but always a fine supply of pollen. When the brood-nest has become sufficiently enlarged it will always be found extending to this south frame, although there may be two or three frames without brood at the north side, and the south frame will be kept throughout the season as a store-house for honey and pollen ; this seeming to be desired by the bees, and convenient, as well, for me. Of course, if the brood-nest is contracted the case is different.   I always, in overhauling, put the frame with most pollen at the south and let the brood-nest commence next to it.

If I do not see the queen on first time going over the frames, which is quite possible, as I do not spend much time looking for her, I look over the frames the second, perhaps the third time, but if not found then, I let her go till some other time. The object in looking for the queen is to see if her wing is clipped. If clipped, I enter in the record-book, (supposing May 10, 1886 to be the date), *May* 10 *q. cl.* (   ) leaving the parenthesis empty. If I find the queen's wing unclipped I clip it and make the entry *May* 10 *cl. q.* (85), meaning that I clipped the queen May 10, and that she was hatched in 1885, although in rare instances I have known a queen to be superseded in the spring.

After clipping the wing of the queen I put her on the top of a frame directly over the brood-nest. If you hold her on your finger over the brood-nest she displays a great degree of perverseness and persists in crawling up your hand, right away from her proper home. So I let her crawl upon a leaf, little stick or other object, lay this on the frames, and she will directly go down into the cluster. If, in overhauling, any frames are found with drone comb, or holes in the comb, they are placed, if containing no brood, at the extreme north side of the frames; if one has brood in, it is placed the north frame of the brood-nest.

As before mentioned, the south frame is broodless, then if there are only one or two frames with brood in, the division-board is put next to them ; if several frames have brood, and the colony is strong, a frame without brood is put north of the brood-nest, before the division-board is put in. The point is to give plenty of room for the queen to lay, but no more than will be actually used till they are likely to be overhauled again.

Generally, instead of returning defective combs, the defects are at once remedied. I turn these combs over to Emma, who mends them at the time. If they are not wired frames she may mend them in this way: She takes a common tea knife with a thin, narrow, sharp blade, cuts out the piece of drone comb if the hole is not already made, lays the frame over a piece of worker comb, (this piece of worker comb may be the part or whole of some old or objectionable comb), with the point of the knife marks out the exact size and shape of the hole, removes the frame, cuts out the piece, and crowds it into the hole. Or, the following plan may be used, and is always used if the frame is wired: After the hole is made, (the mice have probably made the holes in the wired frames), the cells on one side are cut away to the base for a distance of $\frac{1}{8}$ to $\frac{1}{4}$ inch from the hole, and a piece of foundation cut to the right size is placed over the hole and the edge pressed down upon the base that surrounds the hole. The foundation must not be too cold. Before fall these patches cannot be detected, unless by the lighter color where the foundation has been used.

When all the frames are properly placed in the new hive, the now empty one is removed from the stand, and the full one takes its place. Sometimes I make this change when half the frames are in each hive, as they are then easier to lift. The quilt and cover being put on the hive, the old hive is placed in front or a little to one side, with one corner of its alighting-board resting on the alighting-board of the full hive, and in 10 or 15 minutes the few stragglers left in the old hive will have crawled in with the rest of the colony.

In this way the whole apiary is gone over, the hives being cleaned as fast as they are emptied, and then filled up again. The number-tags are removed from the hives at the time of cleaning, and when all colonies are overhauled the numbers are put on the hives in proper numerical order. The same number remains on the same stand through the entire season, and if, for any reason, two hives are exchanged, their number-tags are changed, and the record-book changed accordingly. No duplicate numbers are used, for it would make confusion if there were a No. 1 or No. 2 in the Wilson as well as the home apiary.

You remember, on finding a queen with wing clipped, (and nearly all will be so), I made the entry, *May* 10 *q. cl.* (   ). The full record will appear " No. 1, (231) May 10 q. cl. (   ) br. in 3," meaning there was brood in three frames. The blank in the parenthesis remains to be filled. This may be done at any time when convenient; in the evening, or the first rainy day. Looking at the previous year's record-book I find the queen of No. 231 was hatched in 1884, so in the blank parenthesis I write " 84."

HIVES, COVERS AND STANDS.

Now that the apiary is all in running order, you may want to take a look at it. You " don't think it looks remarkably neat ?" Neither do I. If I had only a dozen colonies and were keeping them for the pleasure of it, I should have their hives painted, perhaps ornamented with scroll work, but please remember that I am keeping them for profit, and I cannot afford anything for looks. Some of them are painted with a cheap, reddish brown, mineral paint, but that was some years ago and they look very dingy. More of them are unpainted, and the oldest of these look dingier still. I suppose they would last longer if painted, but hardly enough longer to pay for the paint. Besides, in the many changes constantly taking place, how do I know that I may not want to throw these aside and adopt a new hive ? I have already changed three times, having begun in 1861 with a full-sized sugar-barrel, changing the next year to Quinby box-hives, then to a movable-frame hive made by J. F. Lester, and afterward when J. Vandervort, the foundation-mill man, came and lived perhaps a year in Marengo, I bought out his stock of hives, and all my hives are now of that pattern. I supposed they were the exact Langstroth pattern, but they have frames $\frac{3}{8}$-inch longer and 7-32 shallower. If I were commencing I think I should have the regular Langstroth size, although mine are so nearly the same, that I don't suppose the result would be noticeably different. They hold

10 frames, and, on some accounts, I should like them to hold 11. They have tight bottoms and no portico.

At the risk of losing caste as a bee-keeper, I am obliged to confess that I never got up " a hive of my own," never even tried to plan one, but I have tried no little to get up a hive-cover to suit me. A hive is so seldom moved that I care less for its weight, but when I, or, more particularly, my female assistants, have to lift covers all day long, when hot and tired, a pound difference in weight is quite an item. The first covers I had for my present hives were 8 inches deep, and I have just weighed one which was a little wet, and it weighs over 18 pounds. I cut down the depth to 4 inches, but they were still heavy, and leaked, no matter how well painted, nor how well seasoned the stuff. I then got up a couple of hundred very shallow and light, covered with white oil-cloth, weighing 4½ lbs. These, when new, are about perfection, but the second season the oil-cloth begins to give way and then they are something of a nuisance. I suppose it might pay to put on fresh oil-cloth every year or two, but I didn't want to fuss so much, and the last I made were covered with tin, being 1⅛-inches deep, the top boards of ⅜-inch stuff, covered with tin, then painted with two coats of white lead. A block 4x⅞x⅞ nailed on one end serves for a handle. This makes a tolerably light cover, 5¼ pounds, is durable and perfectly water-tight. The greatest objection is the cost, from 20 to 25 cents each.

In getting covers, hives and other articles made, I have suffered great annoyance from inaccuracy of work. A cabinet-maker was once quite indignant because I intimated that he could not make a satisfactory bee-hive. I gave him one of Vandervort's make for a pattern, and when I came to see the first one he made, it was a model of workmanship, except that there was no place for a bee to enter and not a frame would go in! (He was not to make any frames.) Of late years I get nearly everything in the flat from G. B. Lewis & Co., Watertown, Wis., and I count myself fortunate in being so near them that freights are not heavy. Their work is remarkable for neatness and accuracy, and they are so upright and accommodating, that it is a pleasure to do business with them. There is a hive-cover of their manufacture, to which Mr. L. Highbarger called my attention, that I might have adopted if I had known about it in time, although heavier than mine. I think it must be perfectly water-tight, even without paint. The top is of two boards, jointed together by means of a strip of tin bent into the shape of an inverted V, and this tin fits into a saw-kerf cut into each board at an angle of 45°.

Some of my stands are very dilapidated affairs. The latest made are cheap and substantial. Three pieces of fence-board, each 2 feet long and 6 inches wide, are nailed upon a cleat at each end, 18x4x1 inch. Two similar cleats, but loose, lie on the ground under the first-mentioned cleats. This makes it equivalent to cleats of two-inch stuff, with the decided advantage that only the loose cleat will rot away by lying on the ground, without spoiling the whole stand. These stands are leveled with a spirit-level before the hives are placed on them, (sometimes not till afterward), being made perfectly level from side to side, and the rear about two inches higher than the front.

The two stands and hives are placed in pairs, Nos. 1 and 2 being 21 inches apart from center to center, Nos. 2 and 3 about 3 ft. 7 in. from center to center, (sometimes more, as my wife prefers them farther apart), Nos. 3 and 4 the same as 1 and 2, and so on alternately. This makes about 3 inches between the two stands that make each pair, and a working space or alley of about two feet between each pair.

This putting in pairs is quite a saving of room; for if room were allowed for working on each side of each hive, only three-fourths the number could be got into the row. But so far as the bees are concerned, it is equivalent to putting in double the number; that is, there is no more danger of a bee going into the wrong hive by mistake, than if only a single hive stood where each pair stands. If hives stood very close together at regular intervals, a bee might by mistake go into the wrong hive, but if a colony of bees is in the habit, as mine generally are, of going into the south end of their entrance, they will never make the mistake of entering at the north end, as you will quickly see if you plug up, alternately, the north and south end of the entrance. When the north end is closed it does not affect the bees at all, but close tho south end, and dire consternation follows. To the bees the pair of hives is much the same as a single hive, and they will not make the mistake of entering the wrong end.

From the back of a hive in one row to the back of a hive in the next row is eight feet, leaving a street about 6 feet wide between the rows.

Trees shade most of the hives at least a part of the day, and at one end of the apiary the trees were so thick that I cut out part of them. I had previously thought that shade was important and that with sufficient shade there was never any danger of bees suffering from heat, but after having combs melt down in a hive so densely shaded by trees that the sun did not shine on it all day long, I changed my mind. I value the shade these trees give, not for the good it does the bees, but for the comfort of the operator working at them. I don't believe bees suffer as much from the hot sun shining directly on the hives, as they do from having the air shut off from them by surrounding objects. I have had combs melt down in hives, tho honey running in a stream on the ground, one of the hives at least being in a shade of trees so dense the sun never shone on it, and I suspect it was for lack of air. A dense growth of corn was directly back of the hives, and a dense growth of young trees and underbrush in front. I didn't know enough to notice this, although when working at the bees my shirt would be as wet as if dipped in the river. I had the young trees thinned out and trimmed up, the corn-ground in grass, so the air could get through, and I now work with more comfort, and no comb has melted down for years, although that may be partly because they are older or wired.

FEEDING MEAL.

I used to read about feeding meal in the spring. I tried it, put out rye meal, and not a bee would touch it; baited them with honey, and if they took the honey, they left the meal. Finally, one day, I saw a bee alight on a dish of flour set in a sunny place. It went at it in a rollicking manner as if delighted. I was more delighted. At last I had in some way got the thing right, and my bees would take meal. The bee loaded up, and lugged off its load, and I waited for it and others to come for more. They didn't come, and that was the first and last load taken that year. I cannot tell now exactly when the change came about, neither do I know that I have done anything different, but I have no trouble now in getting the bees to take bushels of meal. I suppose the simple explanation is that there was plenty of natural pollen for the few bees I had in the first years, but not enough for the larger number of colonies I had later.

About as soon as the bees are set out in the spring, I begin feeding them meal. For this purpose I like shallow boxes, and generally use hive-covers 4 inches deep. These are placed in a sunny place about a foot apart, one end raised three or four inches higher than the other. This may be done by putting a stone under one end, although I generally place them along the edge of a little ditch where no stone is needed, and they can be whirled around as if on a central pivot. One feed-box is used for every 10 to 20 colonies, although I am guided rather by what the bees seem to need, adding more boxes as fast as the ones already given are crowded with bees.

I can hardly tell what I have not used for meal. I have used meal or flour of pretty much all the grains, bran, shorts and all the different feeds used for cows in this noted dairy region, including even the yellow meal brought from glucose factories for cow feed, although, if this last were known, it might be reported that I filled paraffine combs with glucose and sealed them over with a hot butcher-knife. I think this glucose meal is perhaps the poorest feed I have used. As to the rest I hardly know which is best, and I have of late used principally corn and oats ground together, partly because I was using that for horse and cow feed, and partly because I think it may be as good as any.

When the feed-boxes are put in place, in the morning, (and I commence this feeding just as soon as the bees are out of the cellar), I put in each box at the raised end about four to six quarts, (the quantity is not very material), of the feed. The more compact, and the less scattered the feed the better. The bees will gradually dig it down till it is all settled in the lower end of the box, just the same as so much water would settle there. This may take an hour, or it may take six, according to circumstances. As often as they dig it down, I reverse the position of the box, just whirling it around if it stands on the edge of the ditch. This brings the meal again at the raised end of the box. When the bees have it dug down level there is little to be seen on the top except the hulls of the oats, and what fun it is to see the bees burrow in this, sometimes clear out of sight.

It is always a source of amusement to see the bees working on this meal, and the young folks watch them by the half-hour. By night the oats, meal and finer parts of the corn are nearly all worked out, and after the bees have stopped working, the boxes are emptied, piled up, one on top of another, and at the top, one placed upside down so that no dew or rain may affect them. If I think it is not worked out pretty clean, I may let them work it over next day, putting three or four times as much in a box. When the bees are done with it, there will be empty oat-hulls on top, and the coarse part of the corn on the bottom. It does not matter if it is not worked out clean, for it is fed to the horse and cows afterwards.

After the first day's feeding, the boxes must be filled in good season in the morning, or the bees annoy very much by being in the way, and throughout the day, while the bees are at work, if I go among the feed-boxes to turn them, or for any other purpose, I must look sharp where I set my feet, or bees will be killed, as they are quite thick over the ground, brushing the meal off their bodies and packing their loads. Before many days the meal-boxes are deserted for the now plenty natural pollen, although if you watch the bees, as they go laden into the hives, even when working thickest in the boxes, you will see a good many carrying in heavy loads of natural pollen.

It seems to be a beneficent natural law, that bees do not like to crowd one another in their search for pollen or nectar, or else the meal-boxes would be untouched and all the bees would work upon the insufficient supply of pollen. In consequence of this law it is necessary to furnish a sufficient number of boxes, for although the bees will work quite thick if only 5 boxes are left for 150 hives, they will work no thicker if only one box is left.

FEEDING SYRUP.

I have fed barrels of syrup in the open air, and although I have not done so for a year or two, circumstances might possibly arise to make it again advisable. The feed was put in milk-pans and dripping-pans, and at the last I had some tin-pans, purposely made, which were used the rest of the year as milk-pans. They cost about 25 cents each and were made nearly square, being 12 inches long on the bottom and 11 inches wide, 3½-inches deep, and flaring so as to be one inch wider and longer at the top than at the bottom.

After fussing with cheese-cloths and different floats I settled upon the following float : A bottom of boards ¼-inch thick, 11¾ by 10¾ so as to be ¼-inch less than bottom of pan, on which were nailed, or rather the bottom boards were nailed upon, 12 strips 11¾x1¼x⅜, these strips being about ½-inch apart. At one corner, about an inch square was cut out, so that syrup might be poured in while the bees were at work, without pouring directly on the bees.

For filling, I used a common tin watering-pot with the rose taken off, and a funnel with the lower end made about 2½-feet long, so as to avoid stooping.

The syrup is made of granulated sugar. Into a kettle on the stove, I put five quarts of hot water. When at or near the boiling-point, 25 pounds of sugar are slowly poured in, stirring all the while, and the stirring is kept up till the syrup becomes clear. To each quart of this syrup 2 quarts of water are added as it is poured in the feeding-can, using hot and cold water in such proportions that the thinned syrup shall be as hot or hotter than the finger can be borne in it. Even if it should be hot enough to scald bees, it is cooled by what is already in the feed dishes.

There are serious objections to this out-door feeding. You are not sure what portion of it your own bees will get, if other bees are in flying distance. Considerable experience has proved to me that by this method of feeding, the strong colonies get the lion's share, and the weak colonies very little. Moreover, I have seen indications that part of the colonies get none, both of the weak and strong. You are also dependent on the weather, as wet and chilly days may come, when bees cannot fly.

These difficulties are obviated by feeding at night at the entrance, or by feeding in the hive. After trying various feeders, and ways of feeding, at the entrance and in the hive, I have settled upon the old way of feeding in the combs. I think I first got the idea from Quinby's book, or from the writings of L. C. Root.

Filling the combs, with the greatest care that can be taken, is hardly a job suitable to be undertaken in a room carpeted with Brussels. Although I have improved upon my first method of filling, so that the amount of daubiness and stickiness is reduced perhaps fifty per cent.; yet, when Emma, who is the filler, is through filling a hundred frames, although not ordinarily proud, she is decidedly "stuck up." I speak of Emma as the filler, because we carry into practice (although there are only four of us) the doctrine of division of labor, each one having his specialty. I find it a great saving of time. For instance, Charlie puts together the sections, and Emma puts in the foundation, and they are such adepts that it is a pleasure to watch them at work; but if they were to change places, both would make slow, bungling work. As a general rule, I practice each kind of work, until I settle upon what I think a good plan, and then it is given over to the specialist, who may make such improvement as further experience suggests. So, when for the sake of convenience, I speak of doing a certain thing, it may be that I may not have, with my own hands, touched such work for a year.

Now, as to filling combs with syrup: The syrup is made in a wash-boiler, about 4 pounds of sugar to each quart of water. It must be made some little time before used, so as to have time to cool, or the syrup may be made with less water and filled up with cold water. I have a tin-pan $3\frac{1}{2}$ to 4 inches deep, slightly flaring, so as to be about an inch larger at the top than bottom, the bottom about half an inch longer than the top-bar of the frame, and about an inch wider than the depth of the frame. This pan is put in a box without top or bottom, made of 6-inch wide fence-boards, and large enough so that when the pan is in it, there is about an inch play lengthwise, and scarcely any play laterally. Another box is made without top or bottom, 18 or 20 inches deep, the ends of $\frac{7}{8}$-inch stuff, the sides of thin stuff, and the outside dimensions about half an inch less than the inside dimensions of the box first described.

The pan sets on the floor inside the shallow box, and the deep box also sets inside the shallow box, resting on top of the pan. The object of the deep box is, that the syrup, instead of spattering all over the floor and one's clothes, may strike against the side of the box and run down into the pan. An old tin quart fruit-can is placed upside down on a very hot stove, or on the fire, till the end becomes unsoldered and drops off. With a $2\frac{1}{2}$ inch No. 12 wire nail I punch holes in the bottom, making a row around the outer edge about $\frac{3}{4}$ of an inch apart, $\frac{3}{4}$ of an inch inside of this another row,

then inside of this again filling up the space with holes about
¾ of an inch apart. The holes are punched from the inside,
so that the little projections will be outside, which is, I think,
quite important. Near the upper edge two holes are punched
on the opposite sides. Through one of these holes I put a
piece of wire perhaps a foot long, and fasten together the
ends by twisting, then serve the other hole the same way.
Into each of these wires is tied one end of a string, the other
end being fastened to a nail or staple in the ceiling, the two
nails in the ceiling being about 4 feet apart, and the strings
long enough so that the bottom of the can hangs rather less
than a foot above the top of the deep or upper box. The
strings run across and not lengthwise of the box. A boiler
or tub of the warm syrup stands conveniently by, and with
a short-handled 2-quart tin dipper the can is quickly filled,
and through each hole in the bottom of the can runs a stream,
which forces its way into the cells of the frame of brood-
comb, which has been placed in the bottom of the pan. The
can is easily moved about, so as to fill all parts of the comb.

When one side of the comb is filled (no matter if filled to
overflowing), the dipper is slipped under the can, to hold it
out of the way, and one end of the frame is lifted and the
other side of the comb turned uppermost and filled. The
frame is then lifted and put in a super which stands over a
dripping-pan, and when this super is filled, others are piled
on it and filled till the pile is inconveniently high and another
pile is commenced. The hotter the syrup is, the more easily
the combs are filled, but if too hot, you may find that after
standing some time, your nice worker combs have dropped
out, leaving nothing but wires in the frames. About 125° is
as hot as will be safe.

From time to time the pan, in which the frames are placed
to be filled, will become pretty well filled with syrup. I don't
know that it hinders work if it is one-quarter full, but whenever it needs emptying, the deep box is lifted off, then the
shallow one, when the pan can be lifted. I have tried having
the holes in the can closer together, or otherwise differently

made from the plan mentioned, but not with satisfaction. Unless the streams, as they leave the can, continue separate till they strike the comb, the filling will be, to say the least, slower.

When I first had combs filled in this way, much annoyance was caused by the frequent clogging of the holes in the can. Finally, Emma thought of putting a wire-strainer in the top of the can, such a one, I think, as is sometimes used for straining tea, and there has been no further trouble. I have spoken only of feeding sugar, but if I have on hand any dark honey, or that which is in any way objectionable, now is the time to use it.

FURTHER SPRING WORK.

I would like to say that I am very methodical about overhauling and seeing to the building up of colonies, from the time they are placed on the summer stands, till the honey harvest begins, but it would hardly be in accordance with facts. Indeed, I am afraid there have been cases in which a hive has not been overhauled for the first time, till it needed a super. If I were sure of getting around in time to see to each colony the second time, before it had increased much in size, I should always close up with a division-board, leaving barely enough room for the queen to occupy for a short time. As, however, I am not always uniform in the matter, I said little about division-boards when talking about overhauling.

The fact is, every hive has its division-board, which is a very simple affair. A board of inch pine, unplaned, is made about ⅛ inch shorter than the inside length of the hive, and deep enough so that when the division-board is finished it shall reach within ⅜ of an inch of the bottom. This board is re-sawed, making two boards about ⅜ of an inch in thickness. A strip the same length as a top-bar and ⅜x⅜ is nailed on the edge, and the division-board is complete.

It is reasonable to believe that a colony is better able to build up in the spring, if it has only a small space to keep warm, being closed in by a division-board, and well covered over the top, and yet I have been disappointed to find less difference resulting, than it seems to me there ought to be. If my colonies were all good and strong in the spring, I am not sure but I might do as well to give them the full quota of combs and let them fill up at leisure. I usually allow, at the first overhauling, at least plenty of room, although a weak colony may be shut down to the two or three combs in which brood is found.

If I find a colony short of stores, at the first overhauling, it is supplied immediately, either with a comb of honey from some colony which has died, or with a comb of sugar syrup. I have had, one time and another, a good many very weak colonies in the spring, and I am puzzled to know what to do with them. It seems of no use to unite them, for I have united five into one, and the united colony seemed to do no better than one left separate. About all I try to do, is to keep the queen alive till I find some queenless colony with which to unite them.

One year I took the queens of five or six very weak colonies, put them in small cages, and laid the cages on top of the frames, under the quilt, over a strong colony. When I next overhauled this colony, its queen was gone, probably killed by the bees on account of the presence of other queens, but the queens in the cages were in good condition, and became afterward the mothers of fine colonies. I had put two of the queens in one cage, as I was short of cages, and did not attach much value to the queens, and these two did as well as the others. Of course this was an exception to the general rule.

And now my lack of system in spring confronts me, and makes it difficult for me to represent things as they really are. I said the first thing after I got the bees out was to overhaul them, giving stores to such as were short. This is hardly correct as to all times. When I fed in the open air, this feeding rather preceded the overhauling. While the bees were being taken out of the cellar, any specially light colonies were set by themselves or otherwise noted, so that stores might be given them immediately, and generally these were overhauled at the same time. When the general feeding is done in the hives, perhaps I can come nearest the truth to say that feeding and overhauling are simultaneous.

I fancy I can hear my good friends, Mr. A. I. Root and Prof. Cook, saying, " Why don't you keep a smaller number of colonies, so that you can have system enough to be able to tell a straight story, and derive more pleasure and profit ?" I know it would be more pleasure; as to the profit, I doubt. If I had so few that I could at all times do every thing by a perfect system. I am afraid I should have part of the time a good deal of idle time on my hands. Neither is it fair for me to charge my lack of system entirely to the number of colonies. Some of it comes from ignorance in not knowing how to do any better, some of it from changing plans constantly, and perhaps some of it from lack of energy in doing every thing just at the right time.

Whatever may be true about spring feeding, I am pretty fully settled in the belief that it is of first importance that the bees should have an abundant supply of stores, whether such supply be furnished from day to day by the bee-keeper, or stored up by the bees themselves six months or a year previously. Moreover, I believe they build up more rapidly if they have not only enough to use from day to day, but a reserve or visible supply for future use. If a colony comes out of the cellar strong, and with combs full of stores, I have some doubts if I can hasten its building up by any tinkering I can do. So my feeding in spring is to make sure they have abundant stores, rather than for the stimulation of frequent giving. There is a theoretical advantage—how much it may amount to in practice I cannot say—in having the combs filled ; that is, that there is less air-space for the bees to keep warm.

One advantage in feeding in the combs is that your feeders are always ready at no extra expense, for at any time when feeding is needed there are plenty of empty combs, and if the combs are all full there is no need of feeding. Some combs used for feeding may be a quarter or half full of honey, and if I desire this to be removed by the bees, the sealing must be broken before the comb is filled with syrup. I do not know how I like to do this best. If I mash the cells down with the flat of a knife, some of them are missed. An uncapping knife makes good work; but the quickest and easiest way probably, I learned from M. M. Baldridge. Take a common three-tined steel fork with prongs not over one-fifth of an inch apart, and merely scratch the surface. A possible objection to this plan is that the scratching may be too deep, and the cell walls broken down.

In my locality I do not think the colonies can ever become strong and populous too early in the season. Theoretically, at least, then, I see that every colony as soon as it comes out of the cellar, has plenty of stores to last it for some time. I know this is a very indefinite amount. Perhaps I might make it more definite by saying, for an ordinary colony, the equivalent of two full combs of stores. If they have not so much I supply them. I formerly thought it desirable to have any feed given them, as far as possible from the brood-nest, so that they might have the feeling they were accumulating from abroad. Further observation makes me place less confidence in this.

I once gave a frame of feed to a colony having five frames, closed up with a division-board. The north half of the hive was empty. For some reason I put in with the frame of feed a frame of clean empty comb, placing the two combs at the extreme north side of the hive, as far as I could get them from the bees. The bees emptied the frame of feed, and stored it in the empty comb at the side! I think there was plenty of room for the feed in the five frames occupied by the bees. For this spring feeding, then, I put the frames of feed inside the division-board. Its being daubly and sticky is enough to make the bees promptly empty it, no matter where it is placed.

At one time I thought if such a comb were placed in the middle of the brood-nest, they would clean it up and not empty it, but their instinct seems to compel them to empty entirely every cell that is not in proper order. At one time I took the cappings carefully off a part of a finished section, the capping being dark, and put the section back in the super, thinking the bees would immediately cap it afresh with lighter material. Those bees emptied every last drop of honey out of the uncapped cells, and it was some time before they filled and capped them. Of course they may sometimes do the opposite of this. Stimulation between fruit-bloom and white clover is hardly necessary, here.

Any overhauling subsequent to the first, is an easy matter. As a broodless frame was left at the south side at the first overhauling, and the brood-nest commenced with the next frame, I can count that the bees will continue this arrangement, nineteen times out of twenty, if not ninety-nine times out of a hundred. In fact this is just the order that the bees will almost invariably establish of their own accord, the brood-nest commencing in the second frame from the south, and the south frame remaining without brood, but sure to contain a large quantity of bee-bread. So in any examination after the first, I commence at the north side and when I come to the first frame of brood, I need go no further, for I know that the brood-nest will occupy all the rest of the combs except the outside one at the south. If they have not plenty of feed, of course it can be given, although it may not often be necessary to give stores the second time, for in this locality they can get good supplies from fruit-bloom.

I suppose they can forage upon 10,000 fruit-trees without going more than a quarter of a mile. If, however, the first frame of brood I come to, contains only sealed brood, I must look further to see whether they have eggs or very young brood, for it is possible they may have become queenless. If eggs are plentiful, but no unsealed brood, I know that they have a young queen which has commenced laying, and I must find her and clip her wing.

If there is nothing but sealed brood, and no eggs, I am not sure whether they have a queen or not, and it is not safe to give them one till I do know, so I give them, from another colony, a comb containing eggs and young brood. I make a record of giving them this young brood thus : " May 20, no eg g y br," and in perhaps a week I look to see in what condition they are. If I find queen-cells started I am pretty sure they have no queen, and I may let them go on and rear a queen, unless I have one I wish to give them. If it happened that they had a virgin queen when the young brood was given them, the presence of this brood is supposed to stimulate the queen to lay the sooner, and I may find eggs on this later inspection. It may be, however, that I shall find neither eggs nor queen-cell, in which case I consider it probable that they have a queen which has not yet commenced to lay, and they are left for examination perhaps a week later.

This is a good time to salt the ground at and about the entrances of the hives, to kill the grass, although too often I leave it till it has to be cut with a sickle. Grass growing in front of the hive annoys the bees, and that growing at the side annoys the operator, especially if the operator is of the female persuasion, and the grass is wet with dew or rain.

In one case I spoke of leaving a hive to be examined a week later. It is not possible to remember always what is to be done, if as many as 50 colonies are kept; so, in the back of my record-book, I keep a memorandum of work to be done, using arbitrary characters to indicate the particular work required. A bit of the memorandum may be this :

June 13, Mon.   4,   93,   17,   84,   79,
                   o   y   cl   —   ?

June 14, Tues.   172,   184,   139,   148,
                 —   Λ   N   ⌣

In plain English this means that on Monday, June 13, I am to look at No. 4, to see if there are any eggs there ; give a frame containing young brood to No. 93; clip the wing of the queen of 17, which I probably failed to find on a previous day ; put a super on No. 84 ; look at the record to see what is to be done at No. 79 ; on Tuesday 14, destroy the queen-cells in No. 172 ; give a queen to No. 184 ; give a queen-cell to No. 139 ; free the queen which for some reason has been caged in No. 148.

I always keep a lead-pencil tied with a long string to the record-book. No ink must be used in it, for you know it will get a little wet sometimes.

THE HONEY HARVEST.

There are certain things always noticed by a bee-keeper, with much interest, as heralding the beginning of spring or of the honey-harvest. Among these are the singing of frogs, the advent of bluebirds, and the opening of various blossoms. With me the highest interest centers in white clover. As I go back and forth to the Wilson apiary, I am always watching the patches of white clover along the roadside; if your attention has never been called to it, you will be surprised to find how long it is from the time the first blossom may be seen, till clover opens out so bees will work upon it. I usually see a stray blossom days before it seems to have any company. In my location I do not count upon anything usually besides white clover for surplus, so no wonder I am interested in it. Basswood trees are scarce, and I don't know that I ever took any honey in which the basswood flavor could be detected, excepting one year. Raspberries and other sources exist, to be sure, but not in sufficient quantity. If I kept only a few colonies, I might secure surplus from various sources. One or two years I had sections filled with what I believed to be cucumber honey, a pickle factory being in the neighborhood.

Quite likely if a second crop of apple-bloom came a month or two later than the usual time, I might get some surplus from that ; but coming so early I think there are hardly bees enough to store it. Still, the bees are at this time using large quantities of honey for brood, and so the apple-bloom is of very great value. Another advantage is that the great

quantity of bloom has somewhat the effect of prolonging its
time, for the latest blossoms, that with a few trees would
amount to little or nothing, are enough to keep the bees busy.
So it happens that I can scarcely recognize any interim
between fruit-bloom and clover. A few items from a
memorandum for 1882 may be interesting :

Apr. 4.—Last bees taken out of cellar.
May 8.—Plum-bloom out. Bees still work on meal and
sugar syrup.
May 10.—Wild plum, dandelion, cherry, pear, Siberian,
Duchess of Oldenberg.
May 31.—Saw first clover blossom.
June 5.—Apple about done.
June 12.—Commenced giving supers.
June 13.—Clover full bloom—plentiful.
June 20.—Locust out.
Aug. 1.—Clover failing.
Aug. 5.—Robber bees trouble.

You will notice that the earliest apple-bloom (Duchess of
Oldenberg) commenced May 10, while the Janets and other
late bloomers were still in blossom on June 5, several days
after the first clover was seen, making about four weeks of
apple-bloom. Possibly this was unusual—certainly the clover
lasted unusually long, being about 7½ weeks from the time
the bees commenced working on it, for they do not seem to
commence work till after the blossoms have been out some
time. You see that I did not commence putting on supers
till 12 days after I saw the first clover-blossom, and if I had
had only a dozen colonies, I might have waited later, but
with a large number I must commence in time so that all
shall be on as soon as needed. A little time before bees
commence work in supers, little bits of pure, white wax will
be seen stuck on the old comb about the upper part, yet I
hardly wait for this, but go rather by the clover.

Another year (1884), I saw the first clover-blossom on May
21, apple being still in the full bloom ; and I commenced
putting on supers on June 2. One year, I remember, clover
failed on July 4, the earliest I ever remember.

WIDE FRAMES.

The first sections I used were the common, 1-pound, 4¼x4¼ section. I used them in the wide frame, two tiers deep, that is, eight sections in a frame. The supers were exactly the same as the hive, except that there were no entrances nor bottom-boards. One advantage of this was that, at any time, I could change a super to a hive by simply chiseling out an entrance and nailing a couple of boards on the bottom. Or I could use one, at any time, as a hive, without any change, by placing it on a stand and letting the front end project over, *a la* Simplicity. These supers measured 15½ inches, inside width ; and putting into one of them 7 wide frames 2

Wide Frame, holding 8 one-pound Sections.

inches wide, and a dummy ⅜-inch thick, left, theoretically, 1⅛-inches space ; as a matter of fact, it was less than 1 inch. Each frame had nailed upon it two tin separators 3½ inches wide, leaving the open spaces at the top and bottom over the comb-surface ¼ of an inch. The frames, filled with sections, were put into the super, the open side, or side without a separator, being put next to the south side, and the dummy at the north side, then with the chisel I crowded all close together, using no wedges or other means to keep them there. This was a large amount of storage-room—56 sections—to put on at first, but I saw no easy way to avoid it. The bees, of course, were rather slow to occupy so large a space, and as at this time I used 10 brood-frames, these latter were first filled.

Later I came down to 8 brood-frames and used means to bait the bees at once into the sections. In the super I put only 6 wide frames, putting in the center a frame of brood, bees and all, from the brood-chamber, and closing up the remaining 7 brood-frames with a division-board.

The first two wide frames were put in as usual, and the third reversed, so as to present the open side next the brood-comb, then came the brood-comb, and the remaining three wide frames. Thus, the brood-comb had on each side of it eight sections opening toward it, and in most cases the bees would begin to draw out the foundation in all 16 of these sections, within 48, if not 24 hours. If left just in this shape the bees will go on rapidly storing, and begin work on adjoining frames, but there is some danger that the queen may lay in the sections next the brood-frame. Moreover, as soon as a section facing the brood-comb is filled, the bees will commence sealing it with dark cappings, caused. I suppose, by their taking wax from the brood-comb, so conveniently near. So in about a week these frames of sections which are already started, change places with their next neighbors, the fresh ones having their open sides next the brood-comb. A week later the bees will have the work well on, in the four frames, when the brood-comb may be returned below, or used otherwise. Then these four frames of sections in which the bees are at work are alternated with three empty ones, and the bees left to fill the 56 sections.

If, however, as is usually the case, more room is needed, after putting the 7 frames of sections in the super, as just mentioned, a second super is placed above. In this a wide frame is placed at the *north* side of the super, then a brood-comb, then two or three more wide frames and a dummy, the brood-comb serving as a bait, to be moved farther toward the south as fast as not needed at the north. It might be thought that the bees would not readily find their way around outside of the dummy of the first super, up into the second super, but they always did. Even a third super was often gone into. One trouble, connected with this, was that the brood-comb was bulged into the sections, and it was a dauby job to trim it down to its proper size.

Lator, I modified the above plan somewhat. After the first four frames of sections were started, the brood-comb was taken away, and the super filled as follows: At the south side was put a frame of started sections and a frame of empty sections; at the north side was put a frame of empty sections, then a frame of started, then two empty, then one started. The south two frames had the sides without separators facing south, and the north five facing north. The north five were crowded close to the north side. This left quite a space between the two parts into which was put a dummy, and this was crowded to the south side. There are now in the super three frames of started sections and four of empty, and one frame of started sections is left. This is put into a second super above, between two frames of empty sections at the south side, from three to five frames altogether being in this upper super. From time to time the bees are baited along in this, adding frame if necessary until the super is full, then if more room is required arrange this super the same as the first one, and add a third super. When sufficient time had elapsed the lower super was examined for any frames that might be ready to take off.

Another plan was, instead of putting a frame of brood above, to put one or two frames of sections in the brood-chamber of the hive, without any super, putting the sections outside of the brood-nest, separators toward the brood. When the bees started work in these sections, they were put as bait in the super which was then given. This plan had the important advantage of confining the heat to the brood-chamber until the bees fairly commenced storing.

An objection to the use I made of wide frames was the bits of comb and honey between the bottoms of these frames and the tops of the brood-frames. This might be remedied by using the Heddon skeleton honey-board. Another objection was the great amount of labor entailed. For one not over strong it made a great deal of heavy lifting. Yet I secured some good crops of honey by it, never in more satisfactory shape, and I am not sure whether I can do any better by any other system, if I do not take into account the item of labor.

Before I leave the subject of wide frames I will tell how I took the sections out of the frames. It is, I think, easier to take out the whole eight than to take out a single one. I at first laid the wide frame down on the table, face downwards, slipped a little stick about ¼ of an inch thick under each end of the frames, then with a stick pushed down the sections all around as far as the table, then by pulling up on the frame and holding down the sections with the stick, the frame was lifted off, leaving the sections lying on the table. But sometimes, when the sections were unusually loose in the frame, they came to grief by tumbling out prematurely as the frame was being turned over on the table. To remedy this, a board was taken, a couple of inches longer than the frame, and as wide as the depth of the frame. Near each end of the board was nailed a strip about ¼ of an inch thick, the two strips being such a distance apart that when the frame was laid flat upon its face these strips would support the ends of the frame, leaving the sections free to drop the ¼ inch. To use it, stand the frame on the table with the separators toward you; then stand the board on the opposite side against the frame, and lay the two together down upon the table, frame uppermost. If any sections do drop out in turning over, they can only drop upon the board. This is a very good plan if the number to be taken out is not very large.

But when hundreds of frames are to be emptied, every little advantage that will make a difference in speed of performance is worth studying for; so I adopted a plan which allowed them to be taken out very rapidly. This was Charlie's specialty, and he became so expert at it that I think it would be difficult for any one to take out sections faster, no matter what kind of surplus case might be used. At his best he can take out 960 sections per hour. Moreover, I have some doubt if there is any surplus-case used from which the sections are more easily and rapidly taken than from these same wide frames; but I may be very widely mistaken in this, for I have not seen all the kinds. I will now see if I can describe the arrangement used, which is as follows :

Take a super, such as hold the wide frames, and saw it into two parts, so that one part shall hold two wide frames, and have ½ or ¾ of an inch space left; in other words, saw off 4½ to 4¾ inches of the inside width of the super. Nail a bottom on it. Turn the open side of this box toward you, and hang a wide frame in it, on the side toward you, as close to the outside as you can without actually having any part of the wide frame outside the box. Now nail stops in the rabbets tight against the ends of the top-bars, and also nail stops against the lower ends of the side or end pieces of the frame. These stops are to hold the frame securely in its place, when you push against it to push the sections out. It is also better to nail in little wedge-shaped pieces, to prevent the play of the lower part of the frame endwise. Nail upon the bottom a piece the whole length, about two inches wide, and of such thickness that when the sections are pushed out of the frame, they will have but a very little distance to drop upon this piece. Cushion the back for the sections to strike against.

Now for a push-stick : Take a stick of hard wood (pine wears out too soon) about 9 inches long and ¾ of an iuch square. At one end cut a shoulder, clear around, ¼ of an inch deep, leaving the end of the stick ¼ of an inch square and about ¼ of an inch long before reaching the shoulder ; whittle away the other end of the stick somewhat tapering, so that about 2 inches of the end shall be not more than ¼ of an inch thick. Place the box upon the table or bench where you are to operate, having the top tip back 2 or 3 inches more than the bottom, by means of boards nailed under the front edge. It should be very solid. A good plan is to have the table or bench stand against the wall, then the box can rest solid against the wall, having another box at the back of it, if necessary, to bring it further from the wall; then fasten to the table with nails, screws or clamps.

Now hang your frame of sections in the box, separators toward you. If the sections are glued tight in the frame, run a knife between the top-bar and the sections, also between the bottom-bar and the sections; but this is often unnecessary. With the shouldered end of the push-stick, start the sections at each of the four corners of the frame, and sometimes it may be necessary to start them at the middle of the end pieces. You will now find the advantage of the shoulder on the push-stick, for you cannot go more than ¼ of an inch before the shoulder will strike the frame, and sometimes it takes so much force to start the sections, that when the attachment gives away, the sections would be broken if there were no shoulder. Now with the other end of the stick push against the different parts of the sections that lag the most, till they are out of the frame.

In some respects it would be an improvement to have open instead of closed tops, to the wide frames; but I never tried them. I tried some wide frames, holding one tier of sections, and having open tops; that is, top-bars like the bottom-bars, ⅜ of an inch less than the ends of the frames. I liked these better.

HEDDON SUPERS.

After seeing the first super or surplus-case which Mr. James Heddon invented, I resolved to try it. I gave it a pretty thorough trial, putting, I believe, two hundred into use. One advantage I expected from them, was the dispensing with separators. I had seen nice sections of honey secured by Mr. Heddon with no separators; and others, also, had been successful. I was unsuccessful. I do not know, positively, why. It may have been the different management. Mr. Heddon allowed, if he did not encourage, natural swarming. I did all I could to discourage it. I think, compared with mine, that his bees were crowded for surplus room. I think this has a tendency to produce straight combs. At any rate, I failed to produce such sections of honey as would pack satisfactorily for shipping. I found more difficulty than I had anticipated in taking the sectious out of the supers. When first taken from the hive, the sections could be taken out as directed by Mr. Heddon, using a block or follower specially made for the purpose. But after they had remained in the store-room for some time, especially in cool weather, I broke too many sections in taking out, as a result of the necessary fall of some 4 inches.

Moreover, it would sometimes happen, that, on inverting the super, the sections would drop out of their own accord. So, before inverting, I laid a board upon the super, then inverted the two together, having it so arranged that the sections, when pushed by the block or follower, could not fall more than an inch or so. When all four of the rows of sections had been started to the extent of the inch or so, I placed upon them a quadruple follower made by nailing a board across four single followers. The sections, having been already started, would come out without much force ; so, placing my chin upon the top of the quadruple follower, I pulled the super up off the sections, and then lifted away super and follower together, leaving the sections all clear. It was not a very graceful performance, but it was safe and effective.

T-SUPERS.

At the North American Bee-Keepers' Convention at Toronto, in 1883, Mr. D. A. Jones, with his usual hearty manner, was making his Yankee cousins feel at home ; and showed me a super which he recommended for comb honey. I am not sure but he had said something about it, the previous year, at Cincinnati ; but I was not specially interested, and paid little attention to it. Now, however, I thought I could see advantages in it, and upon thorough trial I have adopted it exclusively.

I feel just a little hesitancy in saying much in praise of this super. I like it better than anything I have tried, but it

secms to have found favor with very few others. So little
has been said about it, that I do not even know the name of
of it. Neither do I know who first invented it. I think that
Mr. C. H. Dibbern, of Milan, Ills., invented substantially the
same thing afterward, having probably never heard of it
before. In the *American Bee Journal* for 1884, page 133, he
describes his invention. If any beginner should happen to
read this, and should think of adopting this style of super, I
would say, "Go slow." Although I have tried them and like
them, I am not sure that I know another one, except Mr.
Dibbern, who has said anything, publicly, in favor of them.

For want of a better name they may be called the "**T**-
supers." Mine are made for 4¼x4¼ sections, although they
could very easily be made to take 4¼x5⅝, or 4¼x3⅜, or
4¼x2 13-16. Although mine were first made before seeing
Mr. Dibbern's description, they are rather more like his than
those shown me by Mr. Jones. Quite a number of mine are
made from Heddon supers changed over; but I will describe
those which have been made new.

HOW TO MAKE **T**-SUPERS.

First make a plain pine rim or box without top or bottom,
measuring inside 17⅜x12⅛x4⅝. The lumber is dressed on
all sides, and ⅞ of an inch in thickness. Each piece is halved
at each corner. This is not absolutely necessary, although it
makes a very nice, close, and stiff joint. The outside measure is 19⅛x13⅞x4⅝ inches, so the sides are 19⅛ inches long ;
but halving at the corners makes the end-pieces 13 inches
long, instead of 12⅛, as they would be if not halved at the
corners. At the middle of each end slotted hand-holes are
sawed or cut in. I much prefer these handles or hand-holes
at the ends, although others prefer them at the sides. I get
the stuff in the flat, ready to be nailed together.

To nail them together, I have an arrangement that holds
the two end-pieces up, at just the right distance apart to nail
the side-piece upon them. In a job of two days' work, it
may pay to spend the first day in getting ready, or making

devices to expedite the work. The side-piece is nailed upon the end-piece by driving at each end two 6-penny nails about ¾-inch from each edge. After the sides are nailed on, the super is set on end, and at each end of each end-piece three 4-penny nails are driven, one in the middle, and one a half-inch from each edge. These nails are driven slanting to get a stronger hold, and to prevent splitting the ends of the side-pieces. Common 6-penny and 4-penny nails are used, being stiffer than wire nails, which I use for most purposes.

I now go to the tinner and have him cut me six pieces of Russia sheet-iron for each super; each piece 1⅛x1 inch. These pieces are to be nailed upon the under edges of the sides of the super, so that the centre of the middle piece shall be at the centre of the side, and the centre of the other two pieces half-way between the centre of the super and the inside end. These points are easily obtained in this way: Take a strip of paper just the inside length of the super, fold it double, lengthwise, then double again, and the places of the three folds are the proper places for the middle of each piece of sheet-iron. Each piece is to project inward 5-16 of an inch, and is fastened by two ¾-inch wire nails driven about ¼-inch from the inside edge of the wood. A common shoe-maker's pegging awl makes the holes for the nails. This awl I could not well dispense with. The long way of the sheet-iron piece lies across the grain of the wood. If many are to be made, it pays to make a tool for quickly setting the sheet-iron pieces at the right places. I will tell you how I made mine, which is as follows:

Take a piece of pine 17 5-16x¾x¼ inch, the measurements, except length, being not material. Then make two pieces, each 5x¾x¼; and two more, each 3¼x¾x¼ On the 17 5-16 piece, measure off 3 13-16 inches from one end, and placing one end of one of the 5-inch pieces at this point, letting the other end project over the end of the long piece, nail it there with a couple of ¾-inch wire nails, clinching them. Then leaving a space of 1 1 16 inches, nail on one of the 3¼-inch pieces. In the same way nail the remaining pieces on the other end of the long piece. This tool, when placed in the super close to the side, gives just the right place to nail each sheet-iron piece without the possibility of a mistake. To make sure of the sheet-iron pieces projecting 5-16 of an inch, drive a couple of ¾-inch wire nails partly in, at just 5-16 of an inch from the edge in the spaces where the sheet-iron will rest. For each super two pieces of tin must be cut 13¼x⅝. The cheapest kind of tin will answer for this. One of these pieces is to be nailed on the lower edge of each end of the super, so as to project inward ¼ inch.

To set the piece of tin quickly and surely at the right place, I make the following tool: Make one piece of pine 13x1x¼, and another 12x1½x¼. The only material point in the measurement is that the short piece must be ½-inch wider than the long one. Nail the long piece symmetrically upon the short one, so that the long one shall project ½-inch at each end over the short one, and each side of the short one shall project ¼-inch from under the long one. Placing this tool in the super, close against the end, the strip of tin may be put surely in place. Now, with nine ⅝-inch wire nails, nail it on, putting one nail at each of the four corners of the tin, and distributing the other five nails equally, which leaves them something over 2 inches apart. They should be driven rather more than ⅜ of an inch from the free edge of the tin.

Again the tinner must be called upon. For each super, have six pieces of tin cut 12x1 inch. Bend each piece at right angles, or trough-shaped, its entire length. Two of these pieces put side by side make the shape of the letter T, and are soldered together at the top of the letter T. The pieces of tin being bent in the middle make a support of a half inch for sections, and the upright part being also half an inch gives great stiffness and strength to the support. I thought I would improve the matter and have them still stiffer and stronger, so I made the upright part ⅝ instead of half an inch. I regretted it afterward, for it made too large a space below the separators. A tool to hold the pieces while soldering is made of two pieces of ash wood, each 12x2x¾ fastened together at one end by a common 2-inch hinge. When the pieces of tin are put in this, a spring clamp is slipped over the end without the hinge. These **T** tins, or supporters, must be made of good tin. Mine are made of IX tin. I think it might be an improvement if they were made of one instead of two pieces, but my tinner tells me this cannot be done with ordinary machinery. Now that the **T** super is complete, nothing remains but to place the **T** tins inside the super, on the sheet-iron rests, and fill up with sections.

SECTIONS FOR HONEY.

All the sections I ever used were 4¼x4¼, dovetailed, so I have had no experience with any other. As to thickness or width, I have used mostly what are called pound or 2-inch sections ; although they are in reality 1 15-16. I have also used by the thousand those measuring 7 to the foot or 1 5-7 inches, as well as 8 to the foot, or 1½ inches. I tried a couple of hundred 9 to the foot, and 10 to the foot. I think it not likely that I shall ever meddle again with anything less than 1½ inches wide. One advantage of the **T**-super, as well as the Heddon super, is that different sections as to width can be used in the same super without change, so that it costs but little to make the experiment. If I used no separators I certainly think I should use the 1½-inch sections. With separators I hardly know whether to prefer 6 or 7 to the foot.

As already intimated, I cannot dispense with separators ; yet I have used a mixed arrangement with some degree of satisfaction. The plan is to use both 6's and 8's in the same row. (By 6's and 8's, I mean those that measure 6 and 8 to the foot—in other words, 1 15-16 and 1½ inches wide.) First, two 6's, then four 8's, then one six. Separators are placed between the two 6's, also between the 6's and the 8's. No separators are put between the 8's, and not being at the outside they are generally built in pretty good shape, for it is the outside ones that make the most trouble without separators. Even with separators I have had 6's at the outside built so much one-sided that the lower edge of the foundation has been pushed over and fastened to the separator.

ONE, TWO, AND FOUR-PIECE SECTIONS.

The first sections I ever used were four-piece. At that time there were no others. I never used any except the dovetailed. After the one-piece sections were introduced I tried 500, but did not like them. Later, I tried some two-piece, which I like better than the one-piece. The last lot I got were one-piece, chosen from principle rather than preference. Very few used the four-piece; and so far as I can consistently, I like to encourage uniformity in the matter of supplies. If all bee-keepers used the same description of articles, I think it would result in advantage to manufacturers and consumers. I could be more sure of obtaining promptly an article most common in use. Besides, I thought I could more easily obtain nice, white sections in one-piece. I think the majority of those who buy honey, like the looks of the one-piece best. The one-piece are objectionable on account of the "naughty corner"—a fatal objection when used without separators; they are sometimes out of square, and I never knew one of this sort to stay square with any amount of coaxing; there is also constant danger of breaking. A great point in favor of the one-piece is the ease with which they can be put together. I think Charlie's best speed at making the four-piece was seven in a minute, and with the same ease he puts together twenty of the one-piece in the same time.

SHOP FOR BEE-WORK.

The work of putting together supers, sections and all that sort of thing, is usually done in the winter, or early spring. My shop is 18x24 feet, two-story, with a bee-cellar under it. The upper story is used for storing empty supers, hives and other articles not very heavy, or such as are not often needed. The outside door opens into the middle of the east side of the

house into a store-room; immediately in front of you as you
enter are the stairs leading to the upper story, and at your
right a door opens into the work-room. In this work-room
is a coal-stove, and the room, being ceiled up, is comfortable
in the severest weather.

PUTTING SECTIONS TOGETHER.

The empty supers are brought into the work-room, and, as
fast as filled, stacked up in the store-room, ready to bo put
on the hives. In this work, the first thing is to put the
sections together. As fast as Charlie makes them, he stacks
them up on thin boards, generally using dummies for this
purpose. A dummy being about 18x9, holds 60 sections, if
piled four deep. He breaks very few one-piece sections since
he has had practice, and claims the manner of putting them
together has much to do with it. He grasps the sections in
such a way that the right hand bends a joint at one end, the
left hand at the other, and the middle is bent by both; all
three joints being bent simultaneously, and no one at any
time bent faster than another. The secret of it lies in having
the fingers of each hand on *both* sides of the joint it is bending at the same time. If the section is caught by each hand
at the extreme ends, and those ends brought together, one
joint may be bent entirely before another starts. This not
only makes an extra strain on the one already bent, but each
one must be bent with a quicker movement if bent success-
ively than if bent simultaneously. I think he can make
more rapid work by his plan.

PUTTING STARTERS IN SECTIONS.

As fast as a boardful is made, they are stacked up on his
bench and, as needed, lifted over on Emma's table. This is
provided with a Clark foundation-fastener. If rightly used,
I think the Clark fastener will put in foundation more
securely, more rapidly, and with much less expenditure of
labor than the Parker. The Gray is said to be an improvement on the Clark, but I have not tried it.

To use the Clark fastener successfully, the foundation must be warm; the edge to be fastened quite warm and soft. Common flat-irons were at first used, but they cool too rapidly Common bricks are good, but they break with the heat very soon. So I use two fire-bricks, one to be heating while the other is in use. One of these was broken, but I tied it up with wire, and it is just as good as ever. Having a brick heated, take it out of the fire with a pair of claw-tongs, and put in the other brick. Place the hot brick at a convenient distance in front of you, as you sit before the fastener, and lay the pile of starters between you and the brick. The edges of the foundation next to the brick must be laid even, and when the brick is first taken from the fire see that the pile of starters is not laid close enough to it to melt. If the foundation seems to be getting too soft (although it can hardly be too soft if it does not melt), move it back a little. As the brick gradually cools off, move the foundation closer to it. From 25 to 40 starters should be in the pile at first; enough to reach to the top of the brick, and when only 5 or 10 are left, a fresh pile should be put *under*, as it takes some time for them to warm up.

Putting the section in place, lay the foundation so that the presser will press a very small edge; in fact, I think the less bite you take the better. With a quick motion of the feet, let the presser strike the foundation, *letting the feet fly back instantly*. As you lift the section and turn it upright, the weight of a starter of good size will of itself bring the foundation to a vertical position, although I notice that Emma helps the starter to its place in the act of taking up, by deftly touching with the fingers of both hands, as she turns over the section. By no means follow the instructions usually given, to "draw one side of the section forward a little," " as the presser sinks the foundation into the wood." I think I would rather use the Parker fastener if obliged to follow such instructions. It would make the work slower and more laborious.

Working as I have above directed, Emma puts in 7 starters per minute, at her ordinary rate of working; by hurrying, she has put in 10 per minute.

Since using this machine, the dropping out of starters is of very rare occurrence. Occasionally one has dropped out because the starter was pressed so hard that it was actually cut off, but I think it was because the edge of the presser was a little too sharp. Even then it would only occur when the wax was very soft and pressed very hard.

CUTTING FOUNDATION STARTERS.

I have received foundation in different sized sheets; some of it cut into the proper size for starters, some of it large enough for five, and some as high as ten full-sized starters. I do not like it cut to the smallest size, for I like at least one edge of a starter cut quite true and straight, and to have it so I must cut it myself. After practicing different ways, I have settled upon the one that suits me best. Take a board about 18 inches long and 12 inches wide. On one end nail two or three pieces of section, so that the foundation, when placed upon the board, may not project over the end. At each side drive 1½-inch wire nails partly in, at the proper distances to cut the desired size. A pocket-knife (I use a Barlow), a stick with a straight edge like a ruler, two or more flat-irons, and your paraphernalia is complete.

The room must be very warm, and the foundation must be warm enough so there is no sort of danger of breaking it. Generally the foundation comes in strips of the right width, and needs only to be cut into longths. Take five of the strips and arrange in an even pile; lay the pile on the board, pushing it against the section pieces at the end, and lay another pile beside it. Place the ruler across the piles, against the nails. The hot flat-iron should be before you, supported in some way so the flat side shall be uppermost. Lay the blade of your knife flat upon the flat-iron till it gets hot, hold the ruler firmly to its place and cut across. A little practice will enable you to hold the blade flat against

the ruler, without running away from it or cutting into it.
When all on the board is cut, the pieces are gathered in piles,
40 to 80 in a pile, the edges all made even at one side. These
piles are put on division-boards or dummies till used. The
foundation raises the ruler so high from the board that the
nails are hardly stiff enough. To obviate this, I nail on each
side a strip $\frac{3}{8}$-inch thick, and drive the nails in this.

SIZE OF STARTERS.

I believe in having as little space as possible left unfilled
by the starter in the section. I have tried starters of such
length as to reach from top to bottom, but they sagged and
bulged. I have seen sections that had been completely filled
with foundation, the foundation having been first drawn out
in the brood-chamber; but these sections, when filled, had
an unfinished look about the margin. I settled upon a
margin of $\frac{1}{8}$-inch as the smallest practicable, although it
takes almost too careful work to put in such large starters.
So, in a $4\frac{1}{4}$x$4\frac{1}{4}$ section a starter about $3\frac{3}{4}$ inches square is,
perhaps, the largest admissible.

Years ago my sections were always filled so full by the
bees that they carried very securely in transportation. Afterward I began to have trouble from combs breaking down.
It was due, perhaps, mainly to the bees having too much
surplus room. Some sections would be filled with a nice
comb of honey, not very strongly attached at the top, very
little at the side, and not at all at the bottom. Aside from
depending upon crowding the bees to make them fill the
sections, I wanted a plan whereby I could be sure of having
the sections securely fastened at the bottom as well as at the
top. I tried taking partly-filled sections out of the supers
and reversing them, and even went so far as to invent a
reversible super. I abandoned this, however, and adopted
the plan of putting a starter in the bottom as well as the top
of the section. The problem I had to solve was, how large a
starter I could put in at the bottom and not have it topple
over when warmed up and occupied by the bees. By put-

ting it where the thermometer showed a temperature of 100°
and upward, I felt safe in trying a bottom starter of at least
¾-inch. I mean that the starter measured ¾-inch before
fastened in the section. I put in some thousands of this size
and I am, so far, pleased with them. I have, however, tried
them only one season, and that a very poor one, so that most
of them remain unfilled.

The upper starter was of such size that a space of not more
than ¼-inch was left between the upper and lower starters.
I tried some ½-inch or less, but in some cases, at least, the
bees seemed to think such a little starter had no business
there, and tore it down. A few supers had bottom starters
measuring 1 inch; I hadn't faith enough to try many of so
large size. These, however, worked perfectly well and I
shall hereafter use nothing less, and will experiment further
perhaps, to see how much larger can be used. I do not know
however, that there would be any gain in having the bottom
starter larger.

As fast as the sections are filled with starters they are
piled up on boards, as before, and afterward filled into
supers. The separators are put in at the same time, and the
supers piled up in the store-room till needed to put on the
hives.

There is a feeling of real satisfaction in seeing the larger
part of the store-room filled with piles of supers ready to go
on the hives. How many times I have counted them and
admired the nice even piles reaching to the ceiling! Perhaps I should not appreciate them so much if I had not, years
ago, felt the annoyance of running out of sections or foundation right in the middle of the honey season, waiting days
for it, and the honey wasting. Now, however, I am favored
in being so near to that reliable firm, Thos. G. Newman &
Son, that I can get any thing in the line of supplies in half a
day.

TIN VERSUS WOOD SEPARATORS.

I have used both tin and wood separators in these **T**-supers as also on the wide frames. I used to wonder why some insisted so strongly on the superiority of tin for separators, while others as strongly preferred the wood. Perhaps the difference may be accounted for by inquiring where they are used. From my experience I think I should never want wood separators on wide frames : and I prefer wood for loose separators, as in the **T**-supers.

PUTTING ON SUPERS.

Up to the time of putting on supers, the desire has been to have the bees occupy as many combs as possible. I have had as many as nine frames occupied with brood, without my spreading the brood, or doing anything to urge the bees or queen, further than to see that they had abundant stores. When it comes time to put on supers they are reduced to 4 or 5 frames. The combs that are taken away are sometimes used in making new colonies, and sometimes, if they are not needed elsewhere, they are put in supers, tiered up over other colonies. A colony can thus take care of 40 frames without difficulty. To a very limited extent, I have used them for extracting combs, and I think I might find profit in using more of them in this way.

In shaking the bees off the comb at the time of contracting the brood-chamber, or indeed, when, for any purpose, bees are to be shaken or brushed from brood-combs back into the hive, I have been much annoyed by the behavior of the bees. They seem sometimes to take special delight in running up the sides of the hive and overflowing the top in large numbers, so that great care must be taken in closing up the hive or putting on the supers, or many bees will be killed. This may be avoided by taking out the frames, bees and all, and putting them in an empty hive temporarily, then closing up the hive and shaking off the bees in front, letting them run in at their leisure. But in this case it is unpleasant to have the bees crawling all over the ground in danger of being stepped on, and in danger of climbing up one's feet; there is also some danger of the youngest bees, and sometimes also the queen, not finding their way back into the hive.

To remedy this difficulty, I take an old hive and knock out the front end; then this hive can be placed directly in front of the hive from which the bees were taken, the alighting-board of the empty hive resting its full width on the alighting-board of the full hive, and the bees can be shaken into the empty hive from which they can crawl into the full one. To avoid the jog where one alighting-board rests on the other, I nail a couple of pieces of lath on the inside of the empty hive, letting them project in front about an.inch. These projecting ends can rest on the alighting-board of the full hive, and thus the bees have now a level surface on which to walk right into their hive. To close up the open spaces at the sides between the two hives, a curtain of cotton cloth is attached to the front end of each side of the empty hive, and a single tack, loosely pushed into the full hive, holds the curtain stretched across. The 4 or 5 frames left, are generally put at the south side, a dummy next to them, then a division-board. A super is wider than the space thus occupied; so I put in lath to fill out to the proper width, the lath being cut the length of a top-bar. Before putting on the super, I put on a Heddon skeleton honey-board, which is the same size as the super; the south edge of the honey-board coming flush with the south wall of the hive, and the north edge having an extra top-bar or lath under it.

Adam Grimm once said to me, when I was on a visit to him, that he considered it quite important to have a space for ventilation at the back end of the top of the hive, when surplus receptacles were on. I have used them so ever since, so, when the honey-board is put on, a space is left, at the back end, of $\frac{1}{4}$ to $\frac{3}{8}$ of an inch. The object of this space is to prevent too great heat in the brood-chamber. There is, however, a rather serious objection to this open space : the bees do not work well in the back row of sections, being, I suppose,too cool. I improved upon this,somewhat,by making the honey-boards so that they are closed for about 3 inches of their length at the back end, and if I favored natural swarming, I am not sure but I would dispense with this top ventilation.

It may have occurred to you that the vacant space under the honey-board would be occupied by the bees, and that they would fill it up with comb. It seems as if they would, but they do not; at least not one in a hundred. Perhaps one reason is that they have plenty of room above ; and another, that this vacancy is too open to be warm enough. The hive, being wider than the super, there is an open space, at the side, of 2½ inches.

STARTING BEES IN SECTIONS.

The honey-board being on, the super fits exactly upon it. In order to have the bees commence the sooner in the sections, I put a bait in the super. I take out one of the middle sections, and put in its place a section containing some honey. Sometimes I get this bait by taking from a hive a super that has honey already stored in some or all of its sections, putting in bees and all. There are, however, usually, some sections left over from the previous year, that are partly filled. These make excellent bait. If they are partly sealed, as they usually are, I uncap them, so that the bees may cap them afresh. Almost surely, the bees will at once empty these sections; and just as surely, they will immediately commence to fill them up again. The super is covered over with the quilt or cloth and the hive-cover put over this, Neither of these two are satisfactory. The hive-cover is too large for the super, and the quilt lies down in just the best possible shape to induce the bees to plaster a quantity of propolis all over the sections. I suspect it would be quite an improvement to cover the super with an old-fashioned, cleated, wooden honey-board.

QUILTS OR SHEETS.

I think that Robert Bickford did bee-keepers a great favor by giving them the flexible quilt or sheet, in place of the rigid board, but I have yet to find a sheet or quilt that is entirely satisfactory. I first used them of cotton cloth or sheeting, and when enameled cloth was introduced I felt that the the thing was settled. So it was, but not for a long time. As soon as these enameled sheets became a little old, they would crack and tear, and if, by any means, the bees got to the cloth or cotton side, they made short work of cleaning off the cotton, leaving only the paint. They would find some little place about the edge where they could get to the wrong side, and sometimes, by some means, would find or make holes through the central part.

After many of these enameled sheets were so far gone as to be useless, some of my first quilts of sheeting were still fairly good. They had been filled with paper; and although holes had been gnawed by the bees through the cloth and into the paper, yet in many cases, where they had not gnawed entirely through the paper, the bees covered bee-glue over the gnawed part; and whenever a cloth was well covered with bee-glue, it was sure to last well. I tried the experiment of melting up a lot of bee-glue and painting it on sheeting, but did not succeed. If the sheeting quilts were put on at the time when bees were bringing in propolis most abundantly, and then as soon as the bees had covered all parts, to which they had access, with propolis, the quilts should be shifted so that all parts should be propolized, I suspect such quilts would be quite durable. Indeed, I have practiced somewhat successfully in this direction. In this locality propolis is not abundant till after the harvest has well commenced; so new sheets can, at this time, hardly be put profitably on any but new colonies.

My quilts are now made nearly the same as the first I tried.
Indian Head, or other hard twisted sheeting, is made into a
bag open at one end. Into this is put six or eight thicknesses
of newspaper, so cut or folded as not to come within an inch
of the outside margin of the bag. If the paper is large
enough to fill the bag, the shrinkage of the cloth will curl up
the paper so it will not lie flat. The sheet is then stitched
across through the centre, so as to hold the paper in its place.
The sheet must be large enough to allow for shriukage; I
should think it should be at least an inch too large each way.
Possibly the cloth might be shrunk before makiug, but I am
afraid the bees would gnaw this more.

SHIFTING SUPERS.

After a super has been on long enough for the bees to get
well started, the most advanced sections beiug, perhaps, half
filled, I turn it end for end. This for two reasons: The
ventilation space at the back end has made slower work
there; and there is slower work in the north side of the super
under which there are no brood-combs. I have thought of
trying to remedy this by putting part of the brood-combs at
each side, and filling between with dummies. A few square
inches of comb would have to be in the upper part of each
dummy, so that the queen would go from one side to the
other. I should like brood under the whole super. Heddon's
shallow hive will work nicely, if one has no objection to it
otherwise.

TIERING UP SECTIONS.

When the first super is perhaps half filled, I put an empty
super under it. When the second is well advanced, a third is
added, and more if necessary. This is, however, a matter of
judgment, and cannot be made to conform to strict rules.
If the bees of a strong colony crowd the first super and seem
to be making rapid work, and there is every reason to expect
that they can easily fill a second super, the second is given,
although the first may have very little honey in it. On the

other hand, if the colony is weak, no second super is added
so long as there seems to be no crowding, even if the super
is nearly filled with honey.

Toward the latter part of the honey-flow, when there is a
possibility that it may stop at almost any time, I am more
chary about adding supers. It is better to make sure of
having those finished which have already been given. If it
seems best to give any, they are put on top. This gives
plenty of room if the bees need it, and they are not obliged
to use it unless they do need it.

TAKING OFF SECTIONS.

As fast as supers are filled they are taken off. I do not
think I could be bothered to take off each section as fast as
finished, putting in an empty one to take its place. It would
take too much time. Neither do I like to wait till every
section in a super is entirely finished. Unless the bees are
crowded very much, there will be some uncapped cells in the
outside sections which the bees will be very long in sealing.
If these are waited for, the central sections may lose a little
of their snowy whiteness—the thing which, perhaps, helps
most to sell them.

A super is, then, taken off when all but the outside sections
are finished. This can be pretty well told by glancing over
the top of the super, although sometimes the sections may
be all sealed at the upper part and hardly filled below. A
look at the under part of the upraised super will decide it.
The sharp or wedge end of the chisel is thrust under the
supers to pry apart the attachment of bee-glue.

Unless care is taken, bees will be killed when a super,
which has just been taken off, is put back again. Sometimes
there may be so few bees in the way that the super can be
put on quickly without danger. Oftener too many bees are
in the way for this, so I put one end on its place, and with a
series of rapid up-and-down motions, gradually lower the
other end to its place. This gives the bees time to get out of
the way, and there are seldom any crushed by it.

When the white-clover harvest fails, I take off all supers. I have, however, some years, left on some during cucumber bloom. Sections finished at this time have an unpleasant appearance, as if thinly varnished with bee-glue.

GETTING BEES OUT OF SECTIONS.

When taking off supers, they are smoked over the surface, a little while before removing. This, together with the light, I think, drives down the youngest of the bees—such as would not be able to find their way back to their hives if taken away. If two or more hives are opened at the same time, they have the more time to get down out of the super. Whilst the honey-flow is abundant, I need take no pains for fear of robber bees. After the youngest of the bees have gone down out of the super, all I have to do is to set the super beside the hive, or in any other convenient place; and in from one to three hours every bee will have left it.

Sometimes I prefer to get all or nearly all the bees out of the super before leaving it. In this case I set the super on end on top of the hive, and give it a thorough smoking. This drives all to the side opposite the smoker, and I brush them off while continuing the smoke.

Nothing that I have ever used for brushing bees has suited me so well as the Davis' improved bee-brush—a 15-cent tool made of wire and sea-grass, or some such material. If it gets stuck up with honey, so as to be stiff, it is easily washed out again.

When there is any danger of robber bees—and there is always danger when honey is not coming in freely—the supers are taken directly into the store-room of the shop, and stood on end on the floor, with plenty of room between them. The bees will at their leisure come out of these supers and fly to the light. The room is darkened, all but one place. This is a hole, cut in the south wall something more than a foot square. On the outside, at each side of this hole is nailed a piece of lath some 6 inches longer than the height of the hole, so that the pieces of lath run up some

6 inches higher than tho hole or window. A piece of wire-cloth is stretched across the window and nailed on the lath at each side, and on the wall at the bottom. This makes the wire-cloth run some 6 inches above tho window, there being a space of ⅜ of an inch between tho wire-cloth and the wall of the building at this upper part. The bees on the inside fly to this window, crawl up to the top of the wire-cloth and fly back to their hives ; the robbers never try to get in at the top of the wire-cloth, but always lower down.

I have no shop, only at the home apiary, so elsewhere I have to use other means to get the bees out of supers. By taking time enough each super can be cleaned as fast as taken off. Usually, however, I pile them up on an inverted wooden hive-cover, throw a robber-cloth over the top, and then give them a tremendous smoking from below. Enough aro taken off each trip to make a load home—I am talking now about the final clearing off. This will make 3 or 4 piles of 6 or 8 supers each. Wife generally does most of the smoking of these piles, going from ono to the other, keeping them closed up below, except while blowing in the smoke, and leaving one corner of the robber-cloth open for the bees to escape while she is plying the smoker. This constant blowing the smoker uses up fuel very rapidly, and soon it will be filled with burning coals giving out much heat and but little smoke.

One day when her smoker was in this condition, she broke some small limbs off an apple-tree under which she was working, and breaking them into proper lengths, stuffed them into her smoker. If I had noticed what she was doing, I should have told her it would not burn ; but it did burn, and made the densest kind of a smoke till burned up. Of course under ordinary usage it would go out; but with a hot fire to begin with, and constant blowing, it is just the thing, where a strong and continuous smoke is wanted. The fuel I generally use was recommended to me by the late Mr. Jesse Oatman—rotten apple-wood. The best is where a dead apple-tree is allowed to rot where it stands. A pile of trimmings

of an orchard makes very convenient and good smoker fuel
after it has lain a year or two. For the Clark smoker I am
not sure that I like any thing better than rotten wood. For
the Bingham, I like sound wood—ash and apple are good—
sawed into proper length and split into pieces ¼ to ½ inch or
more square ; only this would be too coarse for the smaller
sized smokers. I like the largest size. One who has used
only the smaller sized smokers can hardly imagine how much
better he would like one that holds a large quantity. I hope
to try for fuel pine shavings as recommended in Mr. Heddon's
book, "Success," and think I shall be pleased.

Instead of having so much trouble getting bees out of
supers, away from home, I think I shall make some arrangement that shall be more nearly automatic. I think I shall
try Mr. A. I. Root's plan of having a tent with a hole in the
top, say 4 inches square. He says if robbers go out of this
hole at the top, they do not know enough to come back in
the same place.

When these supers are taken off before the close of the
honey-flow, the unfinished sections are taken out (I will tell
how hereafter), put into another super, and as soon as the
super is filled with these unfinished sections, it is put back
on a hive for the bees to finish.

HONEY AND FUMIGATING ROOM.

After the unfinished sections are taken out, these supers
containing now only finished sections, are stacked up in the
honey-room. This honey-room is an addition built on to my
dwelling-house. It is 20x15 feet, and the floor timbers are
blocked up with stones, so that the floor will sustain a great
weight without breaking. Five feet are taken off one end,
making a room 15x5 feet for a fumigating room—smoke-
room, we generally call it. On the two opposite sides of this
smoke-room are nailed common 6-inch fence-boards, leaving
a space of about 1½ inches between them. This 1½-inch
space gives a chance to put in movable shelves made of
fence-boards. When sections are to be put on these shelves,

they are taken in on small boards. I happened to have a large number of thin boards about 13 inches long and 4 inches wide; these are mainly used, although larger would be better. Two of the fence-board shelves are put up about 13 inches from centre to centre. The little boards filled with sections are placed with an end at the middle of each shelf-board. As fast as needed, shelves are added till the room is full. Three or four tons of honey can thus be put on these shelves.

FUMIGATING SECTIONS.

Generally, I endeavor to fumigate sections within two weeks after taking them off, and then give them a second smoking two weeks later. Sometimes I have let them go without, and I am not sure that fumigating is necessary, provided no sections contain pollen, and no dead bees are in or about the sections. I am not sure that I have ever seen any mention of it, but I have seen wax-worms in many cases, bred in or upon the bodies of dead bees.

The operation of fumigating is very simple. I fill an old iron-kettle one-third or one-half full of ashes. Into this I put a smaller iron vessel, into which I put a pound of sulphur. I do not like the coarse or roll brimstone, because it is so troublesome to keep burning. The sulphur is put in a compact pile, and a lighted match laid upon it. The door of the smoke-room is then shut, and not opened for 24 hours. A small hole, 3 or 4 inches square, in the door, is covered with glass, through which I can peep to see if the sulphur is burning. I can see no flames, only the smoke arising; an old tin pan being inverted over the kettle, as a safeguard, and also to make the burning slower. I never knew the fire to go out when once fairly started.

Sometimes I fumigate a lot of supers containing sections. The supers are piled alternately in opposite directions, or arranged in some way so that the fumes can freely get to all the sections. If they are in the main part of the the honey-room, I use not less than two pounds of sulphur. I have sometimes gone into the room when it was filled with fumes of sulphur, by filling my lungs and then holding my breath till I came out again. It is best, if possible, to avoid this, especially for those of weak lungs, or those who are nervous.

Sometimes I have had only a few supers to fumigate. In such case I take a wooden hive-cover, put it upside down on the floor, and place a super on it. The super is not large enough to cover the whole of the hive-cover, so one end of the cover is left open. In this end I set a small dish, usually an old oyster-can, containing ashes with sulphur in it. A board is used to cover over this open end of the hive-cover, and to prevent this board from taking fire, a piece of old tin or sheet-iron is placed over the sulphur. Other supers are piled upon the first, and the upper one covered tight. I have been more apt to use too much sulphur in this way than when I smoked a whole roomful. It does no great harm, but makes some of the sections look as if covered with a kind of green mould. I suppose some might say I should burn the sulphur on the top instead of at the bottom of the pile; as the sulphurous acid fumes are heavier than air, hence will fall. I can say, in reply, that my plan has worked well in practice. Moreover, the law of diffusion of gases mixes the fumes with the air, and the fumes rise instead of fall, at least at first, on account of being heated. If you think the fumes too heavy to rise, just put a lighted brimstone match under your nose and try it.

ROBBER CLOTH.

A robber cloth is quickly and easily made. Take a piece of sheeting a yard square or less, and this alone will make a cover to put over a hive or super that will allow no bee to enter. The objection is that it is easily blown off by the wind, and that it can not quickly be put on. To remedy this take two pieces of lath, each about as long as the hive, and lay one upon the other with one edge of the cloth between them. · The cloth is longer than the lath, allowing 6 inches or more of the cloth to project at each end of the lath. Now nail the laths together with 1½-inch wire nails, clinching

them. Serve the opposite edge the same way, and the robber
cloth is complete. You can take hold of the lath with one
hand, lift the cloth from a hive or super, and, with a quick
throw, instantly cover up again your hive or super perfectly
bee-tight. The advantage of being able to handle it so
quickly and with but one hand, is great. During the time
when robber bees are troublesome, I keep 5 or 6 of them constantly in use, and would not be without them, if they cost a
dollar apiece. The cost is not over 10 cents each.

ROBBER BEES.

Of course it is not necessary to say how much care should
be taken to avoid robbing at the close of the honey harvest,
but I think many experienced bee-keepers make the mistake
of taking away anything upon which the robbers may have
been at work, and leaving nothing in its place. If by carelessness I have left a section of honey on a hive, and find the
robbers at work upon it, I can hardly do a worse thing than
to take it away. If I leave it, the bees will stick to it, and
clean it out, and for some time a number of robbers will stick
to it after the honey is all gone, but they stick to that one
spot, and if the empty comb is left there, they keep hunting
it all over and over, and by and by conclude the honey is all
used out of it and go about their business. If the section is
taken away and nothing left in its place, they seem to think
they have made a mistake as to the place and hunt all around
for the missing section, until they force their way into the
nearest hive.

If a weak colony is attacked, I may sometimes take it away,
but if I do, I immediately put in its place an empty hive in
which I put some scraps of comb containing a little honey.
They will rob this out and that will be the end of it.

One time I found a colony at the close of the honey harvest, by some means about at the point of starvation. With
more carelessness than was excusable, I gave them, I think
in the forenoon, two or three combs filled with sugar syrup.
Some time after, I happened to look toward that end of the

apiary and saw what looked like a swarm. The bees had
become excited over their new-found stores; the robber bees
had joined in and the bees seemed to think forage was so
plentiful, that it wasn't worth while to be mean about it,
there was enough for all; so the robbers were doing a land-
office business without let or hindrance. I closed the
entrances of the other hives in the immediate neighborhood,
so that only two or three bees could pass at a time, and then
threw a lot of loose, wet hay at the entrance of the besieged
hive. For some time I kept every thing very wet all around
the hive by pouring on pails of water, and then left them
till next day.

No other hives were attacked. I somewhat expected to
find the queen killed, but she was all right next day, and no
further trouble occurred, as the colony was a strong one, and
when in its right mind, capable of taking care of itself.

I make it a rule to stop operations usually when robbers
are very bad, but sometimes it seems necessary to fight it
out. I have often taken advantage of the plan of making
cross bees or robbers lose themselves, or rather lose the
object they are after by rapidly changing the base of operation. One day at the Wilson apiary I had taken off some
wide frames of sections and wanted to take them from the
place where they were piled up, so as to put them on the
wagon. The robbers were so fierce and persistent that it
seemed impossible to open a crack without their immediately forcing their way in. My wife was provided with a smoker
in full blast, and a big bunch of goldenrod or other weeds.
A robber cloth covered the pile. With one hand I lifted the
cloth and with the other took out a frame of sections, then
quickly dropped the robber cloth in its place, my wife keeping a cloud of smoke in the way of any robbers which should
attempt to enter the pile while the cloth was raised. Instantly the frame was out of the super, the robbers made for
the frame of sections, I made for the wagon and my wife
made for me. Running in a zig-zag, circuitous course, my
wife followed me, puffing and switching at every step, and by

the time we got to the wagon the robbers were lost, the frame was slipped quickly into the super on the wagon, and the robber cloth dropped over it. The Scotch folks at the house had a good laugh over the crazy couple chasing one another through the orchard, but we beat the bees. Under ordinary circumstances it would be better to take an easier plan or wait till dark.

PROTECTION FROM STINGS.

I have been a bee-keeper for twenty-five years, during the last eight of which I have made the production of honey my sole business, and yet I have not reached that point where I care nothing for protection from stings. When I first commenced keeping bees, a sting on my hand was a serious affair, swelling to the shoulder, and troubling fully as much the second day as the first. Now, if I receive a half-dozen stings or more, I cannot tell an hour or two later where I was stung, except as a matter of memory. Yet I think that a sting gives me fully as much pain for the first minute or two now, as it did twenty-five years ago. Sometimes the pain is so severe that it literally makes me groan, especially if no one is within hearing. I sometimes wonder at those who scout at any sort of protection, and query whether there may not be just a little of a spirit of bravado about it. I think I *could* go through a year without any sort of protection, but I do not think I ever shall. A bee inside my clothing makes me very nervous, and I cannot go on in comfort at my work with a feeling of uncertainty as to where and when its little javelin shall pierce my flesh. If I feel it crawling on me, and then cease to feel it because it is on the clothing and not on the skin, I am in momentary dread as to where it shall turn up next; and it is a real relief when it stings me, for I know then the precise spot where it is, and have no further expectations from it.

So I seldom go among the bees without a veil. I may not have it over my face, but it is on the hat, ready to be pulled down at any time. The veil is made of inexpensive material, called by milliners, cape-lace or cape-net. It is 21 inches wide. A piece is cut off as long as the circumference of the brim of a straw hat, and both ends sewed together. Shir a rubber cord in one end of this open bag, thoroughly soak or wash out the starch, and sew the other end on the edge of the hat-brim. Loose ones are made with rubber at both ends. The openings at the wrist and neck of my shirt are small, the cloth lapping over so as to give a bee little chance for entrance. If bees are likely to be on the ground, my pants are put inside my boots, or inside my stockings if I wear shoes. I get a great many stings on my hands, but the inconvenience and discomfort of any sort of gloves would be to me worse than the stings. Mrs. Miller works pretty constantly at the hives the same as myself, and uses no protection for the hands, only for the wrists; while Emma, who has worked less at the hives, likes two pairs of kid gloves, loose, and one drawn over the other. I think as she handles frames more she will discard gloves.

I like to get a sting out of my skin as soon as possible, if not too busy. A little trick in this direction is, I think, not known to all bee-keepers. I am not sure whether I learned it by instinct, or from the writings of G. M. Doolittle. If a bee stings my hand, I instantly strike the hand with much force upon my leg, with a sort of quick, wiping motion. This mashes the bee generally, and rubs out the sting at the same time.

If one thinks of the thousands or millions of bees in a large apiary, it will be seen that comparatively few bees make any attack. Sometimes a single bee will threaten and scold me by the hour, perhaps finally stinging me by getting into my hair or whiskers, and for ought I know the same bee may keep up the same thing for days—I mean the scolding, not the stinging. It is sometimes worth while to get rid of the annoyance by stepping to one side and knocking it down with a stick by a few rapid strokes back and forth in front of my face. I often mash it by slapping my hands together.

Sometimes the bees have seemed very cross, and a little observation has shown these bees to proceed from a particular part of the apiary, and really from only one hive. A careless observer might have said all the bees in the apiary were cross. I have had a few colonies so cross that merely walking by the hive was the signal for a general onslaught. Truth obliges me to say that I have sometimes been so badly stung by one of these, when working at them, that I have taken refuge in inglorious flight, glad to get a respite and scrape out the stings. Just why there should be one or two of these in a year in such marked contrast with others I cannot say. The only remedy I had was to kill the queen.

DRESS FOR THE HOTTEST WEATHER.

During the principal part of the honey-flow, a prominent element of hardship is the endurance of the heat. Sometimes the heat really has made me sick, so that in spite of a press of work, I have been obliged to give up and lie down for an hour or more. At such times you may be sure I am not very warmly clad. One straw hat and veil, 1 cotton shirt, 1 pair cotton overalls, 1 pair cotton socks and 1 pair shoes, comprise my entire wearing apparel. Before noon, shirt and pants are both thoroughly wet with perspiration. In this heated condition, I sponge myself off with cold water before dinner, put on dry pants and shirt, and hang up the wet ones in the sun to be put on next day. I am sure that by this refreshing change, I am able to do more work. It might be thought that applying cold water all over the body when every part is dripping with perspiration might make me take cold. I have never found it so, even if followed up every day. The body is so thoroughly heated that it easily resists the shock, and a brisk rubbing leaves one in a fine glow.

SWARMING.

If I were to meet a man perfect in the entire science and
art of bee-keeping, and were allowed from him an answer to
just one question, I would hesitate somewhat whether to ask
him about swarming or wintering. I think, however, I
would finally ask for the best and easiest way to prevent
swarming, for one who is anxious to secure the largest crop
of comb honey. There are localities where a large crop of
honey is secured in the fall, and in such place, or in any place
where the honey-flow is long enough, a larger crop may be
secured by increase, but I am not so sure about that. If a
man in such a place starts in the spring with 75 colonies, he
may get a larger crop by increasing early enough to 150,
supposing 150 colonies to be the largest number his field will
bear ; but would he not have a still larger crop if he had the
150 all through the season and made no increase ? However
that may be, in my locality, which is not one of the very good
ones, and where it happens that somewhere from the first
week in July to the same time in August, the harvest ceases,
after which the bees hardly get enough to keep themselves—
in such a place I am satisfied that more honey can be
harvested by commencing in the spring with the largest
number the field will bear, and holding at that number,
always providing that the means taken to keep down increase
shall in no wise interfere with the best work on the part of
the bees.

If I were working for extracted honey, I suppose the
matter might be managed, to a great extent, if not to the
fullest extent, by simply giving abundance of room in every
direction ; but with comb honey, I do not believe that an
abundance of room in the brood-nest is compatible with the
largest yield of surplus. I thought at one time that if a
queen had four frames in the brood-chamber the addition of
four more frames could make no possible difference, providing the added four were filled with honey when given. I

believe, however, that it does make a difference. If honey
is stored in any part of the brood-chamber, and all the space
is not needed for brood, the bees seem to get in the way of
thinking that there is the proper place for storing, and
possibly through habit continue storing, even after it
encroaches on the room needed by the queen. This results
in two evils, viz: The nice, white honey that we waut in
the sections, is stored in brood-combs; and the bees are
pretty sure to swarm. On the other hand, if only enough
room is left to barely meet the wants of the queen, that seems
to be left entirely for her, and the combs are filled with brood
clear to the outside. On this account I reduce the number
of brood-combs to four or five when putting on supers. I
confess that I do not feel entirely sure of my ground, but am,
as yet, only feeling my way. Neither do I feel sure that I
have as few swarms as when I kept ten brood-combs in the
hive.

It may be that in the future our breeders will give us a
variety of bees that under fair treatment will never swarm.
But that time is not yet here.

MANAGEMENT OF SWARMING COLONIES.

From my first using movable frames, I think I have kept
my queens' wings clipped, so my experience in having
natural swarms with flying queens has been very limited.
But my experience in having swarms issue where and when
I did not want them, has been very large. Only extreme
modesty and humility prevents my being very proud of so
large an experieuce. If I should ever reach that point where
I shall be equally successful in preveuting swarms, I make
no promise to be either modest or humble.

So long as success in prevention of swarms has not been
reached, it remains an important matter to know the best
thing to do when swarms do issue. Under ordinary circum-
tances some one must be on hand to watch for swarms.
With as many as 100 colonies in an apiary, the one who is on
watch can hardly be allowed to do anything else. The

regular noise is so great among so many that the added noise
of a swarm is hardly noticed ; so sight, not hearing, must be
depended on. I have gone on with my regular work and
taken a look once in five or ten minutes along the rows to see
if any swarms were out, but it is not a very satisfactory way
of doing. A bright boy or girl can watch very well, if faithful. It is not necessary, of course, to watch all day ; and the
weather has much to do with the hours at which swarms may
be expected. On a hot morning a swarm may issue as early
as six o'clock, but this is exceptional ; and if the weather has
been cloudy through the day, clearing off bright and warm
in the after part, a swarm may issue after four o'clock.
Ordinarily, however, it is not necessary to be on the lookout
before 8 a.m., or much after 2 p.m. I had a swarm issue once
in a shower, but that is so unlikely to occur that I would not
think it worth while to keep any watch at such a time.

The watcher will soon learn the points of advantage from
which he can easily command a view of the whole apiary,
not needing to stir from his seat unless a swarm issues.
Sometimes, however, there is so much playing going on
among the bees, that there is no alternative but to travel
about and take a close look at each colony that shows
unusual excitement. It is an advantage at this time to have
the hives in long rows. I have 40 hives in a row. Three
such rows, 8 feet apart, make 120. At the middle is a shady
place to sit. A clock or watch lies in open sight so that a
look at every hive may be taken once in five minutes. If
there is no time-piece to go by, the watcher may become
interested in something else, and think the five minutes not
up when double that time has passed; but having the time
measured out, he is free to read or do anything else between
times. At each five minutes, the watcher, who is sitting at
the middle of the middle row, rises, glances along the back
row to the north end ; then, along the middle row to
the north end ; then, stepping forward, glances along the
front row to the north end ; then along the same row to
the south end ; then to the south end of the middle]row ; and

lastly to the south end of the back row. All this has taken less time than it takes to write it, and the watcher is ready to sit down till another five minutes is up.

If, however, unusual commotion is seen—and, sighting along the rows in this way it can easily be seen—the watcher goes to the hive for a closer look. Up to the middle of the day or later, there is not often much excitement, unless there be a swarm; but after this time so many colonies take their play spells that the watcher needs to spend most of his time on his feet.

The watcher is provided with a number of queen-cages. These are easily made and the material costs less than a cent apiece. I take a pine block, 5x1x½-inch, and wrap around it a piece of wire-cloth 4 inches square. The wire-cloth is allowed to project at one end of the block a half-inch. The four sides of this projecting end are bent down upon the end of the stick and hammered down tight into place. A piece of fine wire about 10 inches long is wrapped around the wire-cloth, about an inch from the open end, which will be about the middle of the stick, and the ends of the wire twisted together. I then pull out the block, trim off the corners of the end a little so that it will easily enter the cage, slide the stick in and out of the cage a number of times so that it will work easily, and the thing is complete. When not in use the block is pushed clear in, so as to preserve the shape of the cage. Such cages can be carried in the pocket without danger of being injured.

When the watcher finds a swarm issuing, he is pretty dull if he does not become interested in looking for the queen. I do not know of any sure way to find the queen, but she is not often missed. I think I can find her most easily by watching on the ground in front of the entrance. Very frequently she comes out at the back end of the hive at the place left for ventilation. Rarely she may be found at some distance from the hive, on the ground, with a group of bees about her. If not found, she is most likely in the hive, and the swarm may re-issue in a day or two. She may be lost,

but at this particular time her loss is not so very great.
There is no danger of the swarm being lost; it will return to
the hive in a few minutes, although I have known them to
cluster for half an hour or more before returning. It may
happen, sometimes, that a swarm may go into a hive whose
colony has swarmed a little while before, and where it is
always peacefully received. I do not like this doubling
up, but I do not know that I lose anything by it, for the bees
can store up just as much in ono hive as another.

When the watcher finds the queen, she is caged. Either
the cage is held down for her to run into, or she is allowed to
run up on the finger and then caged. After the queen is in
the cage, the block is pushed in an inch or so, and the cage
put where the bees can take care of it, usually in the vacant
part of the brood-chamber, which is accessible without taking off the super. The number of the hive is taken.

A few years ago Mr. G. M. Doolittle gave a plan for
management of swarming colonies when no increase was
desired. I do not know that he uses it now. I do not know
that I shall ever use it again, and yet it was valuable to me,
and for some circumstances nothing may be better. The
plan, in brief, was this: The queen being caged and left in
the hive, all queen-cells are cut out in five days from the time
the swarm issued, and five days later all queen-cells are
again cut out and the queen set at liberty.

I used this one season with great satisfaction, and I do not
remember that any colony thus treated swarmed again.

The next season I varied the plan. Instead of leaving the
queen with the colony to remain idle for ten days, I took
her away and gave her to a nucleus, a new colony, or where-
ever a queen was needed. At the end of the ten days I
returned her to the colony, placing her directly upon a comb
taken from the middle of the brood-nest. Often, however, I
gave them a different queen, for after an absence of ten days,
I doubt if they could tell their own queen from any other.
Besides, they were in a condition to take any queen without
grumbling.

After the first year, however, I had some colonies swarm again after the queen was given them. Whether it was the season, the change in the plan, or some other cause, I am unable to say.

I then adopted a plan which relieved me of the necessity of hunting for and cutting out queen-cells. No matter how careful I might be, there was always a possibility that I might overlook a queen-cell, although this very rarely happened, if ever. But it took a good deal of valuable time. I give herewith the plan, which I think an improvement :

When a swarm issues and returns, it is ready for treatment immediately ; although usually it is put down in my memorandum of work to be done, and the time set for it may be the next day or any time within five days, just as suits my convenience. The queen is caged at the time of swarming, and put in the vacant part of the brood-chamber—possibly in the upper part of a super—where the bees can care for her.

Within the five days, I take off the super, and put most of the brood-combs into an empty hive. Indeed I may take all the brood-combs, for I want in this hive all the combs the colony should have. In the hive left on the stand, I leave or put from one to three frames, generally two. These combs must be sure to have no queen-cells, and may be most safely taken from a young or weak colony having no inclination to swarm. The two combs are put in the south side of the hive, a division-board and dummy next to them, and the supers again put on. If I did not do so at the time of taking out the frames, I now shake off the bees from about half the frames, not being particular to shake them off clean. This hive is then put on the top of the supers, the queen let free on top of the frames, and the hive covered up. A plenty of bees will be left to care for the brood, the queen will commence laying, all thought of swarming is given up, and every queen-cell torn down by the bees. In perhaps two days I take a peep to see if the queen is laying, for it sometimes happens that at the time when I " put up the queen, " as I call the operation I have just described, there is already

a young queen just hatched, and then the old queen is pretty
sure to be destroyed. In this latter case I may remove the
young queen and give them a laying one, or I may let the
young queen remain.

In ten days from the time the swarm issued—sometimes
ten days from the time I "put up the queen"—I put down
the queen. If, by chance, a young queen is in the upper
hive, I do not like to put her down until she commences laying and her wing is clipped, for fear of her taking out a
swarm. It seems a foolish operation for them to swarm
when there is nothing in the hive from which a queen can
be reared, but I have had it happen. The operation of putting down is very simple. I lift the hive off the top, place it
on the ground, remove the supers, take the hive off
the stand, place it on one side, put the hive containing the queen on the stand, and replace the supers. At the
time I put up the queen I changed the number-tag, so as to
keep the number always on the hive containing the queen.

You will see that this leaves the queen full chance to lay
from the minute she is uncaged, and at the time of putting
down there will be as much brood as if the queen had
remained in her usual place. Most of the bees, of course,
adhered to the lower hive when the queen was put up, but
by the time she is put down quite a force has hatched out,
and these have marked the upper hive as their location. Upon
this being taken away, the bees, as they return from the
field, will settle upon the cover, where their hive was, and
form a cluster there; finally an explorer will crawl down to
the entrance of the hive below, and a line of march in that
direction will be established immediately. In a day or two
they will go straight to the proper entrance.

We left, standing on the ground, the hive with its two
combs, which had been taken from the stand. These two
combs, when the queen was put up, probably had a good
quantity of eggs, and brood in all stages. They now contain
none but sealed brood, some queen-cells and a pretty heavy
supply of pollen. Or, it may be that eggs from an imported

queen were given, and the queen-cells are to be saved. A goodly number of bees adhere to the two combs and I know of no nicer way to start a new colony, than simply to place the hive in a new location. Or, the bees may be shaken off at the old stand and the combs used again to do duty as they have done during the last ten days, or given to a nucleus which needs them.

It may be objected that this keeping bees queenless for ten days makes them work with less vigor. I am not sure but it ought, but I must confess I have had no strong proof of it come directly under my own observation. So far as I could tell, these bees seemed to work just as hard when their queen was taken away as before. In the spring of 1885 one colony was, by some means, left entirely away from the proper rows—some three rods from any other colony. I took it away, put it in proper line, and left to catch the returning stragglers a hive containing one comb, this comb having no brood and very little if any honey. This colony having been a very weak one, very few bees returned to the old spot, but these few surprised me by filling a good stock of honey in empty comb, before they were put with the rest of the colony.

Swarms treated on this " putting up " plan often swarmed again, but if they did they were put up again. An objection to the plan was that these " put-ups " were in the way and had to be lifted down when anything was done with supers. Still, for any one who allows the bees to swarm, and who does not object to the lifting, the plan is a good one.

PREVENTION OF SWARMING.

I have, however, been anxious to find some way by which I could *prevent* swarms. I am not sure that it can ever be profitably done, but I am not willing to abandon the effort without a faithful trial.

If I knew all about just what makes a colony swarm, I would be in better shape to use preventive measures ; but I do not know. Of course there are some general things that I know about it, such as heat and want of room in the brood-nest, but a good deal of mystery envelops it. I have tried taking away the queen of a colony which had swarmed, and giving in its place a queen which had just commenced to lay, leaving no queen-cells in the hive. Within three days the young laying queen would come off with a swarm.

One season I kept eight brood-combs in the hive, and every week or ten days took out two of the central combs, replacing them with foundation or empty combs. This was to give the queen so much room that there should be no desire to swarm. It was successful in most cases, but there were too many exceptions to make the plan reliable.

The next season I settled upon a plan which I felt pretty sure would prevent the possibility of swarming. It was a no less radical measure than to keep the colony queenless. I reasoned that as I had never had a queen hatched inside of eleven days from the time the queen was taken away, or from the time the bees started queen-cells, the colony was safe from swarming if once in ten days I took away their brood and gave them fresh; also, that it was only bees over two weeks old that worked in the field ; add to this the three weeks that it took from the egg to the full-fledged worker, and it was five weeks or more from the time the egg was laid till the bee became a gatherer. Clearly, then, only such bees as came from eggs laid five weeks or more before the close of the honey harvest were available as gatherers. Why not have the colony queenless during this five weeks ? So I took away the queen, leaving in the hive three combs, one of which contained eggs and brood in all stages, the other two containing nothing from which queen-cells could be started.

Once in ten days the comb of young brood with its queen-cells was taken away and a fresh one given them, and at the close of the five weeks, which was about the close of the harvest, the queen was returned. As a preventive of swarming, it was a complete success. Not one colony thus treated swarmed ; how could they ? As a means of securing a large crop, I think it was an egregious failure ; although I can hardly tell with great definiteness, the season itself being a failure. Possibly the absence of the queen itself had something to do with lessening their stores, but I doubt it. But when all combs of brood but one were taken away, a large force of prospective bees were taken away that would have hatched out in from 1 to 21 days. All the brood taken away when the queen was taken, even the eggs, would have produced bees in time for at least part of the honey harvest. Besides, those that might have hatched during the last week of the harvest, although they might gather not a drop, would be able to take the places of older ones, that in the absence of recruits were obliged to stay in the hives to do home duty.

If I had allowed four or five frames of brood, changing every ten days, the result might have been quite different. Moreover, the one frame they did have was, for the most part, filled with brood so young, that little or none of it hatched while in the hive. If I should try anything in the same line again, I should keep four or five frames in the hive, and this should be mainly brood well advanced so that much of it would hatch out to replenish the wasting numbers.

The problem, however, which I am most anxious to solve is, how to manage to have no swarms, and still allow the queen to remain laying in the hive all the time. It may never be solved, but it is worth some dreaming over.

QUEEN-REARING.

My sole business being to produce honey, I am not particular to keep a popular breed of bees, only so far as their popularity comes from their good qualities as honey-producers. I am anxious to have those that are good gatherers, good winterers, not cross, and not given to much swarming. I have no great confidence in my ability as a scientific breeder, so I have not attempted to establish a strain of my own, but every year or two I send to A. I. Root for one of his best imported queens. I know that there is more or less of uncertainty about this way of doing, but I am not sure that I know any better way. I am at least sure of good stock, and I am equally sure that amongst my own rearing I have bad some very bad stock. I get the queen in July if I can, more likely in August. I prefer this time because a queen then costs less, because it is less trouble to establish the queen in a colony of her own than early or late in the season, and because such queens are reared at a season of the year when they are more likely to be good. She is of little or no use to me till the succeeding year, but she is in good condition for that.

I do not mean to be understood that all my bees are pure Italians. Some of them are very fine; excelling in beauty any I ever reared from imported queens ; and some have a good share of black blood in them. Neither do I mean that all my queens are reared from imported stock. When it is convenient I often rear a queen of what appears good stock, and frequently the bees take the matter into their own hands and supersede their queen with one to suit their own notions. I have had good queens at three and four years old, but as a rule I suspect better results might be had not to keep them so long. A queen, however, which seems to be doing good work, so long as she remains quietly in her place, is in no great danger of decapitation, for it is quite a disturbance of the domestic arrangements to change the queen of a colony that is bending right down to solid business ; but if, by swarming or otherwise, a queen is out of her colony, her chances of getting back alive are not very good, if she is over two years old.

It is of great importance to have good queens, and in former years I did not hesitate to break up my strongest colonies to secure good conditions for queen-rearing, but I found—perhaps I ought to say stumbled upon—a better way.

I do not want my imported queen to do very heavy laying, as she will last longer if not overworked; and I prefer to have a rather weak force of bees with her during the busy season, letting the colony build up somewhat late, so as to be in good condition for winter ; so through the main part of the season her hive stands on the top of the hive or super of some other colony, changing to a new one as often as it contains too many bees. Among other things, this prevents her from swarming. When her hive is changed to a new place, the field force readily unites with the colony over which it has previously stood.

About the time the honey-flow fairly commences, I make preparations for queen-rearing. The first thing wanted is some worker comb, preferably new, evenly filled with eggs. I take one of the middle combs of the hive containing the imported queen, and fit centrally into it two pieces of worker comb taken from one-pound sections. These are about four inches square and I select those which have been drawn out about the proper depth for brood-rearing, or trim them to that depth. The honey has all been removed, probably the previous year. A piece is cut out of the brood-comb for each section and the section merely crowded in. I do not mean, of course, any of the wood of the section, just the comb. Between the two holes cut out for the sections to be crowded in, about an inch space is left. If a wired frame is used, the wires must be cut. If I remember rightly, Mr. A. I. Root objected to my mutilating combs in this way, but as the holes are immediately filled up, no mutilation appears.

Suppose these section combs thus prepared to be put in the middle of the brood-nest on June 1, I look, on June 2, to see if eggs are to be found. Most likely; if not, almost surely June 3. About 3 days from the time eggs are first laid, I cut out these sections and replace them with fresh ones. Then the sections are cut up and attached to brood-combs in the manner directed by Mr. Alley in his book on queen-rearing, only instead of leaving an egg in each alternate cell, I leave one in every third cell. Scraping off the top-bar of the frame which contains these eggs, I write on it " 16, " as June 16 is the earliest day possible on which a queen can hatch. I know it is considered that it should take 16 days for a queen to hatch, but in very strong colonies I have known them to hatch in 15. This frame is then given to a strong colony, usually one that has swarmed. Of course the queen must be taken away. It is put between two other frames, neither of which has any eggs or unsealed brood, although it is better if it has sealed brood. These three brood-frames are the only ones in the hive, and all the bees have to do, is to rear queen-cells and fill their supers. I have spoken of one hive, but the two sections ought to furnish eggs enough for at least three or four colonies to start queen-cells from. If they are colonies that have just swarmed, there is no loss, for they would be kept queenless anyhow.

I can hardly think of any way possible by which I could rear stronger or better queens; but among them I have reared some not worth keeping. Bees seem to have an instinct for starting a succession of queen-cells, and where a natural swarm has issued there may be several days between the hatching of the first and last queen-cells. Of course these are started from eggs or grubs of different ages. In the present case, the bees probably start what they think the proper number for a beginning, and a few days later when they want to start more there are nothing but well advanced larvæ, so they use them. The result is poorly developed queens. To avoid this, all larvæ not in incipient queen-cells should be destroyed in 24 or 48 hours after the eggs have been given.

I have always used the regular-sized frames for queen-rearing nuclei, but I do not usually occupy a whole hive for a single nucleus. Years ago I separated a hive into six apartments having one frame each; an entrance at the front end, one at the back, and two at each side, one of which was at the front lower corner and one at the back upper corner. This worked well, but was open to the objection that a nucleus hive could be used for nothing else. As I do not make a business of rearing queens for other than my own use, I prefer to use regular hives with very little changing; so I put a division-board in the middle and have two nuclei in a hive. The middle part of the entrance is closed and each nucleus has its own end of the entrance. There is more likelihood of a bee or queen going into another hive than of going into the entrance at the wrong end. The greatest trouble is to have

the division-board so tight that no bee can go from one
side to the other.

As I make much use of these double hives, I am not satisfied with any division-board that I have ever used. Although
they may be all right at first, I have lost a good many queens
by finding that later there was a passage from one side to the
other under the division-board. The only way I can account
for it is, that in the hive it is always dry and if any change
takes place in the division-board, at all, it shrinks, although
made of thoroughly seasoned stuff: on the other hand, the
body of the hive being exposed to the weather may swell, especially toward fall. Either the shrinking of the division-
board, or the swelling of the body of the hive, or both
together would cause a passage under the division-board.
I am thinking of remedying this, by nailing in the bottom of
the hive and at each end, a little tin trough in which the
division-board, if it shrinks, may play up and down without
danger of making any passage for bees.

As I have previously described, the division-board is a plain
board about ⅜-inch thick, having a top-bar ⅜x⅜. This top-
bar is nailed down at each end by a ¾-inch wire nail. Over
the whole hive is now laid a piece of cotton cloth or sheeting
that will fully cover it and allow for shrinkage. Over this
cloth, directly upon the top-bar, is laid a piece ⅜x⅜, exactly
like the top-bar, and nailed down near each end with ¾-inch
wire nails. When the hive is filled there is no possibility of
any bee crossing from one side to the other at the top, until
holes have been gnawed through the cloth on both sides.
My hives being 10-frame ones, there is room in each side of
these double hives for four frames and a dummy. I have
sometimes used five frames in each side, but it makes them
too crowded.

Having these nucleus or double hives thus arranged, a
nucleus is started in each of them. In each is put two frames of
brood and bees, next to the division-board, so that each
nucleus may have the benefit of the heat of the other. As I
have already intimated, there is no nicer way to establish

nuclei than to take bees of a swarming colony. Most of them will stay wherever they are put. In fact, any bees that know they are queenless, are more apt to remain attached to their combs wherever they are put.

As the queens may hatch out ou June 16, it is well to have these nuclei made two or three days in advance, and it is absolutely necessary to do so if the bees for the nucleus are taken from a colony having a queen. They must be queenless long enough to fully realize their condition or they will tear down the queen-cells given them. I have, however, started the nuclei frequently just at the time of cutting out and giving them queen-cells, by taking a good number of bees from a colony or larger nucleus engaged already in building queen-cells. If I want to use a nucleus or a pair of them, for queen-rearing only, to be broken up afterward, I generally put them on the top, over some other colony, so that the bees may be there united when the nucleus is taken away.

For cutting out queen-cells or for cutting comb for any purpose, it is quite important to have the right kind of knife. For some time, I used the small thin blade of a pocket-knife, but it was too short; the large blade was too thick. I happened upon an old-fashioned case-knife or tea-knife that suits me well. It is of fine steel, extremely thin, the blade 4½-inches long, and ¾-inch wide at the widest part and ½-inch wide toward the point.

On June 1, the empty comb was given to the imported queen, on June 4 or 5 the eggs were given to the strong, queenless colony, and on June 15 or 16 the queen-cells must be cut out and given to the nuclei. I have had queen-cells hatch out in good shape by simply placing them between the combs, and even by laying them on top of the frames, but I think I have had better success by the old-fashioned way of patching them into the comb, right among the brood.

In some cases I have done well in giving virgin queens, caged for a day or two, but my experience has been limited. About five days after giving the queen-cell, on June 20 or 21.

I take a look at it. It may have hatched, the bees may have torn it down, or it may remain unhatched. If unhatched on June 22, it is likely dead, and if still alive, I have my doubts about the value of a queen having no more ambition than to be so long in hatching. If the cell shows that the queen has hatched, but I do not happen to see her, I enter on the record "q. h.," meaning queen hatched. If I see the young queen I make the entry "s. y. q.,"—saw young queen. In either case, I give the nucleus a frame containing some eggs or young brood. If the cell is torn down or bad, I give them another.

My experience has been that a queen only a day or two old is as easily found as an old queen; but as she grows older, up to the time she commences laying she is very hard to find. I almost always see the queen the first time I look, after she has been hatched; but I do not know that it is any better than to see the cell properly opened. About a week later, on June 28, I look to see if the queen is laying.

If I find eggs, I clip both wings of the queen on one side; generally the two left wings. I use a pair of small embroidery scissors. I catch the queen by the wings, with the thumb and finger of the right hand, and while clipping her I hold her by the thorax with the thumb and finger of the left hand.

If I find no eggs, and no queen-cells are started, I feel confident they have a queen, and look a few days later to see if she is laying. If there are no eggs, and I find queen-cells started, I suspect they are queenless, the queen having probably been lost on her bridal tour. I am not sure of this, however, for sometimes they will start queen-cells while a virgin queen is in the hive, but if by July 1 there are still no eggs, and the queen-cells not torn down, I consider it a pretty sure thing that it is a hopeless case.

In the record-book something like the following entries may be found: " June 16, g q c (16—21.) 21st s y q, g y br. 28th cl q. " This means that on June 16, I gave the nucleus a queen-cell that could not hatch before June 16, and ought

not to hatch later than June 21 ; that on June 21 I saw the
young queen and gave them young brood ; and that on June
28, I clipped the wing of the queen which I found laying.

INTRODUCING QUEENS.

If I wish to take one of these queens to introduce elsewhere, my favorite plan of introduction,—almost, indeed,
the only one I ever used,—is to place the queen, without cag-
ing, directly on the middle of a comb of brood, right among
the bees. I must first be sure that these bees have started
queen-cells. If I wish to introduce an imported, or other
valuable queen, I give her to a frame of brood with
young bees just gnawing out, or else to bees that I am sure
are young, by having moved them a day or two before, in the
middle of the day, to a new location, so that the old bees
might all fly back to the old location.

I gave my general way of introducing a queen,—by that I
mean when I introduce the queen alone, but generally it
happens that I can take the frame of brood on which I find
her, and put it, bees and all, where I want the queen to be.
In time of a honey-flow, I have even done this without
being sure that they realized their queenlessness.

INCREASE BY NUCLEI.

After a nucleus is started, it is an easy thing to build it up
into a colony, by adding frames of comb or foundation as
fast as the bees can take care of them. During the hot
season of the year, I do not hesitate to spread the brood of
these young colonies, for there is little danger of chilling, and
I think they can be made to rear brood faster than if left to
themselves. One year I took 12 colonies to the out apiary,
increased them to 81, and took 1,200 pounds of honey, by
means of this nucleus process. All the help they had from
other colonies was the eggs from which to rear queens; but
they had no combs to build, being furnished with ready-
built combs, and the honey taken was extracted buckwheat.

I do not know that there ever was such a yield of buckwheat here before or since.

UNITING NUCLEI.

There is much irregularity in the matter of rearing queens. Sometimes every queen seems to know what is expected of her, and commences laying right along ; while at other times many fail. Perhaps both queens in one double hive do nicely ; in another, both fail. It may be the first queen-cell given fails, and a second, and even a third, may meet the same fate. I do not want to keep on rearing queens after the harvest is over ; so I must make allowances for these failures, by starting more than I want.

Suppose toward the close of the season I have used all the queens I want in full colonies, the nuclei that I did not care to keep have been broken up, and there remain some queenless among the nuclei in the double hives intended to be built up into full colonies. If the nucleus in one side of a double hive is queenless, and the other has a laying queen, the two are united. A passage is made in some way through or under the division-board, and in two or three days nearly all the bees will be found in the side having the queen. The combs, or a sufficient number of them, may then be put together. Sometimes I make the union at once by simply putting the division-board to one side.

It may be that No. 5 has a queen in each side, while No. 6 has one in neither. In this case I take from No. 5 the frame containing the queen and put it, bees and all, in No. 6. Both sides of No. 6 may be at once united, and the same thing may be done at No. 5 a couple of days later. In this way every hive contains at least one queen, and some of them two. These latter will remain as two separate colonies, unless by means of a defective division-board they take the matter of uniting into their own hands.

FALL FEEDING.

If colonies have not enough stores for winter at the close of the clover harvest—and with my present management they are not likely to have—I can hardly expect them to gather from later sources more than will meet their daily needs, till winter closes in upon them. So they must be fed; and there is nothing, so far as I know, to be gained, and perhaps something to be lost, by postponing the feeding till late. In my locality, perhaps there is no better time than August. Of course, in some localities, it would be folly to feed so early, if indeed it would be necessary to feed at all.

The feeders are the same as used in the spring—combs—and the paraphernalia for filling the combs the same as that described in spring management. The syrup is made stronger—5 quarts of water to 25 pounds of sugar; perhaps I had better say 5 pounds of sugar to one quart of water. I think that a stronger syrup might be better, as requiring less evaporation by the bees, but 5 pounds to the quart is about as thick as can well be filled into the combs without having it too hot for safety. An even tea-spoonful of tartaric acid for every 20 pounds of sugar is stirred into the syrup about the time the sugar is dissolved. The tartaric acid is first dissolved in a little water. I am not sure whether the tartaric acid is necessary, although if not fed to the bees the syrup without the acid will granulate in a short time. It takes a long time for the syrup to cool enough to be poured into the combs, and it is well to have a tub of syrup left over from the previous day to mix with the hot.

Care must be taken to let no bees get in where the combs are being filled, and still greater care that none which get in shall get out alive. Every precaution must be taken to avoid getting the bees started to robbing. The entrances of the hives should be closed all but two or three inches. If the entrances were shallower it might make it too warm to close up so much. The combs must be put into the hives as near dark in the evening as convenient. I have put them in as late as 9 o'clock, using a lantern.

Generally, I put two frames of feed at a time in each hive, outside the division-board, being sure that there is a passage for the bees to get to it. The large size of my hives comes very convenient now, for there is plenty of space to easily put in the frames even in the dark, there being only four or five frames occupied by the bees. Of course I have previously made an estimate as to how many frames of feed each colony will need, by looking them over. Some may need none, and some be entirely destitute. These latter will get about four frames of feed (given on two different evenings), and others in proportion. There is little danger of giving too much. In the double hives the colonies are not large where two are in a hive, and three combs is all they need for the winter. There is room for but one frame of feed at a time, outside the dummy; and this frame is the first one reached by any bee coming in at the entrance. This makes it less secure against robbing, but I want the two nuclei to be in the middle of the hive for warmth, and they are not to be disturbed again before winter. As the feeding is done in the evening, I have had no difficulty from robbing.

The same set of combs is used over and over again—there is no trouble in taking out the empty combs in a day or two by daylight—and when all feeding is done these combs are put away in supers in the shop, securely covered up from mice. I know of no way in which mice can more rapidly do a bee-keeper damage, than by getting at empty combs, especially those in which brood has been reared.

SHAKING BEES OFF COMBS.

Speaking of taking the empty combs out of the hive, reminds me of a little device worth knowing, in getting bees off combs that are not too heavy. Hold the frame in the left hand, by one end of the top-bar, and with the closed fist of the right hand pound on the back of the left hand. A very few strokes will take off the last bee. Let the strokes be made in pairs, first a light stroke, and then very quickly after it a heavy one. If the combs are heavy, this plan will not work. For heavy combs I know of no plan better than one I learned from G. M. Doolittle, which is this:

Take hold with each hand at the ends of the top-bar, supporting the weight by the first and second fingers. Raise the frame, letting the thumbs be well raised from the top-bar. Now let the frame fall, and as it falls strike hard upon the top-bar with the ball of each thumb. This gives the frame a jerk, and it immediately gets another jerk by being suddenly stopped from further falling, by the fingers. There is something about this having the jerks in pairs, that breaks the hold of the bees upon the combs in a way that cannot be done if more time is allowed to elapse between the jerks.

DOUBLE HIVES FOR FULL COLONIES.

After the bees are fed they are ready for winter quarters, without further preparation. There is, however, one thing that I have practiced to a very limited extent, and from my experience so far, I expect to practice it more in the future. I mean putting two colonies into one hive. From the time the bees are fed in the summer or fall, till perhaps the middle of May, most of my colonies would have room enough in one-half of a ten-frame hive. I am not sure that any of them ever need more room through the fall and winter, and in the spring they need no more till more than four frames are needed for brood. With some, this may come quite early, but I think I should be well satisfied if I could get all my colonies to contain four combs well filled with brood by the middle of May. Some of them may have at that time brood in 9 or 10 frames, but more of them could have all their brood crowded into three or four combs.

Now, if during the time I have mentioned, we can have two colonies in one hive, we shall, I think, find it advantageous in more than one direction. It is a common thing for bee-keepers to unite two weak colonies in the fall. Suppose a bee-keeper has two colonies in the fall, each occupying two combs. He unites them so they will winter better. If they would not quarrel and would stay wherever they were put, he could place the two frames of the one hive beside the two frames in the other hive, and the thing would be done. Now suppose that a thin division-board were placed between the two sets of combs, would he not see the same result? Not quite, I think, but nearly so. They would hardly be so warm as without the division-board, but nearly so; and both queens would be saved. In the spring it is desirable to keep the bees warm. If two colonies are in one hive, with a thin division-board between them, they will be much warmer than if in separate hives. The same thing is true in winter. I have had weak nuclei with two combs come through in good condition during a winter in which I lost heavily; these nuclei having no extra care or protection other than being in a double hive.

Now suppose we have 100 colonies that are all fed up for winter and they are then put into double hives. Please understand that there is little or no extra expense for these double hives. They are just the regular hives and we have the division-boards anyway; only we take special pains to see that the division-board is perfectly bee-tight. If the hives are to be hauled home, as I haul mine each fall, there are only 50 instead of 100 to haul; just half the bulk, and a much less weight than the 100 would be. Just half the hives are to be handled in taking in and out of winter quarters; just half the room is occupied in winter quarters; and I think, although I do not know, that the bees will winter better than if only one colony in a hive. If they are to be taken, in the spring, to a distant apiary, there is the advantage of hauling only 50 hives instead of 100. If, in the spring, any colony be found queenless it is in fine position to be united with its fellow colony.

CHANGING FROM SINGLE TO DOUBLE HIVES.

Possibly you may be ready to agree with me so far as to say, "Certainly, the thing looks desirable, but is it feasible? Will not the trouble counterbalance all advantage?" I know it is usually a matter of some trouble to change a colony from one location to another in the same apiary. I think, however, that I have reduced the trouble to a minimum. I will give you my plan and you can judge for yourself.

As I have already told you, my hives stand in pairs, and I kept them so, years before I thought of double hives. Suppose there are 100 colonies in the apiary and we want to put them into 50 hives. All are fed up ready for winter, and each one has four combs. I am not sure that all colonies can be reduced to four combs, as I have never reduced all of my colonies to this number, and I have sometimes wished my hives were 11-frame instead of 10-frame, so that my double hives would hold five frames on each side. I might, however, have the division-board a little to one side, and have five frames on one-half the hive, and four in the other. I have spoken heretofore of keeping the brood-frames on the south side of the hive. This has been my general custom, but I have practiced to some extent having the entrances of each pair of hives at opposite ends. For instance: Nos. 1 and 2 stand close together. The hives facing east, the brood-nest of No. 1 is at the north side, and of No. 2 at the south side. This is perhaps the better way. The bees of No. 1 use mostly the north end of their entrance and the bees of No. 2 use the south end. When the bees are fed, only these ends of the entrances are left open.

Now remove Nos. 1 and 2 from their stands, and remove one of the stands and put the other in the middle of the space occupied by two stands. On this stand place a double hive prepared as already described for queen-rearing nuclei. Put the combs and bees from No. 1 in the north side of the double hive, and those from No. 2 in the south side, and cover up the double hive. A few bees will remain in the old hive, and these may be placed in front at each side of the double hive, the alighting-board of the old hive resting on one corner of the alighting-board of the double hive. In a short time all the bees will have crawled into the new hive, when the old ones may be removed. Put the number-tags from the old hives, each on the proper side of the front of the double hive.

The matter is now accomplished and it has been no long or difficult job. The bees use the new entrance *almost* as readily as the old. To them their hive seems moved less than its width to one side, and there is no possible danger of their entering the wrong place. I have tried it, and watched the result, therefore I speak of not what the bees *ought* to do, but what they *do* do.

CHANGING FROM DOUBLE TO SINGLE HIVES.

Can we as easily get them back into two hives in the spring when they become crowded in this double hive? Just exactly as easily. We simply reverse the operation. Take the double hive from its place and replace it with the two stands and two hives, then remove the contents of the double hive and put them in the proper single hives, and the bees will go every time to the right place. I speak again from personal observation as to what the bees actually do.

ENTRANCES OF DOUBLE HIVES.

I am not sure just what is the best size for the entrances of these double hives. They are not used in hot weather except for nuclei, and I have done as follows:

I first put in a wooden plug to close up that part of the entrance in front of the division-board. This plug is about 1x⅞x½ inch, and generally a little lump of beeswax is used to wedge it tightly in place. The plug being of the same

width as the thickness of the hive stuff, ⅛ of an inch, it
comes flush with the front surface of the hive. I then take
a piece of lath 8 or 9 inches long, set it up on its edge against
the middle part of the entrance, lay against the lath a 1-inch
wire nail, and crowd the nail into the bottom or alighting-
board by means of the thumb-nail or a chisel.

MACHINE FOR EMPTYING T-SUPERS.

My arrangement for taking sections out of the T-supers is
made in a common wooden hive-cover, 8 inches deep, 21
inches long, and 17½ inches wide, inside measure. These
exact dimensions are not absolutely essential, but I happened
to have this on hand. Practical difficulties, however, would
come in the way of anything much smaller. Now make a
box without top or bottom, 16½x11, and 6 inches deep. The
hive-cover being set upside down, put this smaller box in it,
near one corner, so that the side of the box shall be 1 7-16
inches from the side of the hive-cover, and the end of the
box 1 5-16 inches from the end of the hive-cover. Now nail
the box securely to the hive-cover, and cut out that part of
the hive-cover which is now the bottom of the smaller box.
This whole arrangement, so far, is merely to hold a bearing-
board in place.

The bearing-board is now to be made. Take a board 16⅝
inches long and 11 inches wide. Take boards 12 inches long
and ¼-inch thick and nail them across the first board so as
to just cover its length, and project ½-inch at each side.
This makes a surface 16⅝x12 inches. If this bearing-board
be now put inside an empty T-super, and the T-super raised,
it will be seen that the bearing-board will easily drop through
the super, except where it is upheld by the three pieces of
sheet-iron on each side. Places must be cut out of the
bearing-board so that the sheet-iron pieces will present no
hindrance. In order to make these places abundantly large,
I cut them 1½x½ inch. When cut out, the measure will be,
from the corner of the board to the first place or hole, 3¼

inches, then 1½ inches for the hole, then 2 13-16 to the next hole, and the same from each corner.

Now place this bearing-board on the top of the box in the hive-cover, just where a super will most easily pass down over it, having one corner of the super all the while held tightly up in the corner of the hive-cover. Nail the bearing-board there with two or three nails, leaving the heads of the nails out, so that they can be afterward drawn out with a claw-hammer. If your work has been well done, when a super is placed over this bearing-board, the corner of the super being held tight in the corner of the hive-cover, there will be an equal space at each side between the bearing-board and the side of the super, also at each end between the bearing-board and the end of the super.

Now turn over the hive-cover. The part of the hive-cover which was cut away allows access to the bottom of the bearing-board. Two stops are to be nailed on the bearing-board, so that when not nailed to the box it can be quickly pushed to the exact place it now occupies. One stop 7½x ¾x½ (neither of these dimensions is essential), is placed across the bearing-board, tight against the end of the box, one end of the stop being tight up in that corner of the box nearest the corner of the hive-cover. The other stop, about 1 foot long, is to be placed lengthwise of the bearing-board, tight against that side of the box which is nearest the hive-cover. Now turn over the whole affair and draw the nails that fastened the bearing-board to the box. By reason of the stops, the bearing-board can be placed upon the box and instantly pushed up to its place.

If you now attempt to lift the bearing-board, having the fingers of each hand under each end, you will find there is not room enough for the fingers of the left hand, the end of the hive-cover being in the way. A place, therefore, 6 inches long and 3 inches deep is cut out to make room for the hand, and enough is whittled away on the under part of the bearing-board at each end, and also at the top edge of the box to allow plenty of room for the fingers easily to get a hold under the end of the bearing-board. The two stops prevent the bearing-board from standing level and solid on a table, so another stop or small block must be nailed on the unsupported part toward the corner, but near enough the centre not to interfere with its free working. The under part of the bearing-board, which is of ⅞ or 1 inch stuff, is better in two pieces or split through in the middle to prevent warping. Pains must be taken to see that the hive-cover used is perfectly square.

TAKING SECTIONS OUT OF T-SUPERS.

To take out sections with this arrangement, I place it in front of me on a table—no fastening is necessary—so that the box inside the hive-cover shall be nearest to that side of the hive-cover which is next to me, and the end of the box which comes nearest the end of the hive-cover shall be at my left hand. The bearing-board is now put in place, and pushed tight in the left hand corner. The super full of sections is placed on the bearing-board and crowded close to the left hand corner. I now lean forward, throwing the weight of my body partly upon the super, and pressing with the left fore-arm upon the end and opposite side. Then with the closed fist of the right hand I strike upon the farther corner of the super at the right hand. This breaks the attachments of the sections at this corner, and I then strike upon the different parts of the super so as to get it started all around. Then putting a hand on each end of the super, I push it evenly down and let it drop in the hive-cover. The bearing-board is lifted out with its load of sections, and the now empty super is also lifted out.

It is often better, perhaps always, to run a case-knife around so as to cut through the propolis that may fasten the upper part of the sections to the super. The fist will become sore if used for much pounding, so I use a *heavy* hatchet or hand-axe. With this it is not necessary to strike heavily, whereas a light hatchet must be struck so hard that it would mar the super and not start the sections so easily. It is important to bear down upon the super while striking.

The above is for the late work when all supers are to be emptied. You will remember that the first supers that were taken off had the unfinished sections put back on the hives to be finished. I will now explain how this was done.

The sections may all be taken out of the supers, or the finished ones may be left in the super till later, merely taking out the unfinished ones. To accomplish this latter, blocks must be put in the bottom of the hive-cover—I mean the inside bottom as it lies ready for use—so that the super cannot fall clear down, but will fall upon the blocks and then be only so far down that $\frac{1}{2}$ or $\frac{3}{4}$ inch of the sections are still in the super. The unfinished sections can now be picked off the sides, for they will always be found in that part of the super. Then I lay a board about the same size as a bearing-board upon the sections, and hold this board down with my chin, while I raise the super by the hand-holes at each end. As soon as it is raised enough for the thumbs to reach on the top of the board while the fingers still remain in the hand-holes, I relieve the chin from further duty, and raise the super up till the sections are fully in place, holding the board down by pressing with the thumbs. The super is then piled up till the regular time for taking out all.

It is well to have two or more bearing-boards so that they can be taken directly to the scrapers, and the one who takes out, can be taking out on one board while the scraper is emptying another. One bearing-board can be made to do, by having several plain boards of the same size or a little larger. Lay a plain board on the table and the bearing-board full of sections beside it, letting the projecting half-inch of the thin boards of the bearing-board rest on the edge of the plain board. Now slide the sections in a body from one board to the other.

SCRAPING SECTIONS.

When the sections are taken out they go directly to the scraper. Mrs. M. and Emma always do the scraping, and a first requisite seems to be to array themselves in the most hideous apparel. At least the outside garment must be of that description, for bee-glue flies in all directions when they are scraping, and a coating of bee-glue on a Sunday gown is no great improvement. A common case-knife with a straight blade is the tool used for scraping. A seat is used 6 or 8 inches higher than a common chair. Generally a common wooden chair is set on a wooden hive-cover. A little box or block 6 or 8 inches in length and width, and perhaps 2 inches thick, is placed on the table, and the section put on this block to be scraped. All four sides of the sections are scraped clean of propolis, and the edges as well. It is not a difficult job for a careful hand, but a very disagreeable one. The fine dust of the bee-glue is very unpleasant to breathe. A scraper should be a careful person, or in ten minutes time he will do more damage than his day's work is worth. Even a careful person seems to need to spoil at least one section, before taking the care necessary to avoid injuring others. But when the knife makes an ugly gash in the face of a beautiful white section of honey, that settles it that care will be taken afterward.

PACKING SECTIONS IN SHIPPING-CRATES.

The scraper has in easy reach two shipping-crates. In one, all perfect sections are put as fast as scraped. In the other are put any which are a little off color, either as to comb or honey, or which have some cells unsealed. The most difficult thing about the packing is to prevent veneering. It seems to come so natural, when a particularly white and straight section goes into the crate, to put it next the glass, best side out at that. But it is especially desirable that the outside shall be a fair index of the entire contents of the crate. In the long run there is money to be made by it, to say nothing of the feeling of satisfaction.

Any sections which are not enough filled to go into the second-class crate, are set to one side to be extracted, unless some of them are saved to be used as bait in supers during the next season. The sections to be extracted are put in wide frames without separators, uncapped, and extracted in a Peabody extractor. If I had much extracting to do, I

Shipping-Crate for one tier of Sections.

should get a better extractor, and I suppose I might have something better than a wide frame to hold the sections, but I have so little extracting to do that I use just what I happen to have.

After these unfinished sections are emptied of honey, I get the bees to clean them. One way is to put a lot of them on a hive; another is to pile them up in supers out-doors, covering them up and leaving a hole only large enough for one or two bees to pass at a time. If they were left entirely open, the sections would be torn to pieces, and possibly robbing started; but I have never known any harm to result where only one or two bees were allowed to pass at a time. When cleaned by the bees, these sections are filled into supers and piled up in the shop ready for the next season.

I have used, generally, shipping-crates holding 24 one-pound sections, the sections being two tiers high in the crate. I have used some holding only a single tier. These latter bear transportation better, but they cost more per pound of honey and hardly present so good an appearance.

KEEPING HONEY.

I have sold a crop of honey before it was all off the hives, but latterly I have not generally sold the last of it till spring. It is not the easiest thing in the world to keep it through the winter in good shape. If kept cold it is apt to granulate or candy, as it is usually called. If allowed to freeze, the combs crack and look bad, and in time the honey

Shipping Crate for two tiers of Sections.

oozes out of the cracks. Honey is deliquescent, absorbing from the atmosphere a large amount of water if conditions are favorable. Try putting some common salt in a place where you think of keeping honey : if the salt remains dry, so would honey. But a place that is suitable at one time may not be at another. One year I filled my smoke-room with honey. It was a good place for it; the outside walls were thin and the heat of the sun made it a hot place. When cold weather came, however, it was a bad place, and the lower sections at the back part—beautiful, snowy-white, when first put in—became watery and dark-looking. A fire for cooking was kept in the adjoining room, and although there seemed but very little steam in the air, by the time it got to the back end of the smoke-room, and settled to the lower part, there was enough to spoil hundreds of sections. You see, warm air is like sponge to take up moisture, and cold squeezes the moisture out of it. The point to see to, then, is to have no air coming from a warmer place to the place where the honey is. I would sooner risk honey in a kitchen with a hot fire and plenty of steam, than in a room without fire and with a door partly opened into a sitting room where no water or steam is ever kept. Indeed, a kitchen is quite a good place to keep honey.

If comb honey became granulated or watery, I know of no way to restore it. If for home use, or if one happens to have a market where extracted honey sells for a good price, the sections may be put in stone crocks, *slowly* melted, being sure it is not overheated, and then when cool, the cake of wax may be lifted off the honey.

The best place to keep comb honey is also the best place to keep extracted; but if extracted honey becomes granulated or watery, it may be restored to its former, or even a better condition. If thin and not granulated, by setting it on the reservoir of a cook-stove and letting it remain days enough, it will become thick. I suppose you may have known this, and also that extracted honey, when granulated may be liquefied by slowly heating, but did you know that when thin honey is warmed for a long time the flavor is improved ? I have had the flavor improved and could attribute it to nothing but remaining a couple of weeks on the reservoir. I do not mean by this that if fine-flavored honey in good condition is placed on the stove reservoir it will be improved. Most people, however, who have had much to do with honey, must have noticed that when extracted honey becomes thin from attracting moisture from the atmosphere, it seems to acquire a different flavor,—perhaps I might say it has a sharp taste—and the slow heating seems to restore it partly if not wholly to its former condition. The same thing is true of honey which is taken thin from the hive, not yet having been brought to proper density by the bees.

There is a difference of opinion as to whether honey, or perhaps nectar, evaporated outside of the hive, is equal to that which remains in the hive till thick. Of course, no large amount could be evaporated on a stove reservoir. Some bee-keepers have large tanks in which to evaporate honey by the sun or other heat.

There is another plan which I have used to secure some extra fine honey for private use. Whether it could be used profitably on a large scale, I cannot say. There are, however, always people who are ready to pay a high price for an extra article. After a crock of clover honey has granulated, I turn it on its side or upside down, and let it remain days enough to drain off all the liquid part. If drained long enough, the residue—and this will be nearly all the crockful—will be as dry as sugar, and when this is liquefied by slow heating it makes a delicious article. It will, however, granulate very easily a second time. On a larger scale, the liquid might be drained off by boring a hole at the lower part of a barrel of granulated honey. I spoke of heating clover honey in this way: I do not know what other kinds may be treated the same way, but I have had some granulated honey of smooth, even texture, from which no liquid part could be drained. When set to drain, the whole mass would roll slowly out.

MARKETING HONEY.

I have had no uniform way of marketing honey. I should prefer in all cases to sell the crop outright for cash, if I could get a satisfactory price; but many, if not most years, I can do better to sell on commission. Judgment must be used as to limiting commission-men to a certain price. Some commission-men will sell off promptly at any price offered, and when sending to such men it is best to name a certain figure, below which the honey must not be sold. I have sold in my home market, as well as in towns near by, and have shipped to nine of the principal cities, and it would be an impossibility for me to say what would be my best market next year. Prices vary according to the yield in different parts of the country. If shipping to a distant point in cold weather, I keep up a hot fire to warm the honey 24 hours before shipping. If very cold I wait for a warm spell. On a wagon, the length of a section should run across the wagon— on a car lengthwise of the car. I always prefer, if possible, to load the honey directly into the car myself. Then I know that it will carry well, unless the engine does an unreasonable amount of bumping.

Much has been said about cultivating a home market, but there are two sides to the matter. If bee-keepers from neighboring towns come in and supply my home market at two cents per pound less than my honey nets me when shipped to a distant market, about all I can do is to leave the home market in their hands. I suspect, however, that it would have been to my advantage to have paid more attention to developing my home market for extracted honey.

In deciding between a home and a distant market, there are more things to be taken into consideration than are always thought of. There is breakage in transportation, and the greater the distance the greater the risk. If I can load my honey into a car myself, and it goes to its destination without change of cars, I do not feel very anxious about it. On this account a car-load is safer than a small quantity, for a full car-load may be sent almost any distance without re-shipping. If re-shipped, it is not at all certain how it will be packed in a car. I once sent a lot of honey to Cincinnati, and when it arrived at its destination, the sections were actually lying on their sides! I suppose the railroad hands who packed it in the car at the last change, thought the glass was safest from breaking if the case was put glass side down. The strangest part about it was that I lost nothing by the breakage. The dogged persistence of a German consignee obliged the railroad company to pay all damage; for the consignce was that staunch German and genial friend of bee-keepers—C. F. Muth. It is the only case in which I have known a railroad company to pay for breakage of honey.

There is less danger of breakage by freight than by express. Besides danger of breakage, there is risk of losing in various ways. You may not be able to collect pay for your honey. If sent on commission, the price obtained may be less than the published market report. You have no means generally to know how correct the claims for breakage may be. In fact, unless you know your consignee to be a thoroughly honest man, you are almost entirely at his mercy. A quarter or half a pound may be taken off each case by the claim that it is custom to reject fractions. Taking all these things into consideration, together with the cost of freight and shipping-cases, and it must be a good price that will justify a man to ship off honey to the neglect of his home market. If shipped to be sold on commission, providing he ships to a near market, the price should be at least 2½ cents per pound more than he can get in his home market, to justify his shipping. If he ships to a distant market the difference should be still more, as the additional freight may make a difference of one cent per pound or more, and the risk of breakage becomes greater.

BRINGING BEES HOME IN THE FALL.

In the fall, the bees must be brought home from the out apiary so as to be wintered in the cellar. If I were in a latitude where out-door wintering was safe, I should be glad to leave them the year round without moving. I have studied somewhat upon some plan by which they might be safely left, and am not without hope that sometime cellar wintering may be so perfected that I can build at little expense a cellar for each out apiary where the bees may be left without attention from fall till spring. I have not reached that point yet, so I feel obliged to have all my bees brought home in the fall. There are always a few things upon which bees can work till quite late ; so it is desirable to be as late as possible bringing them home. They must, however, be brought home early enough so they will be sure of a good flight after

being brought home, and before being put into the cellar.
For the past five years ending with 1885, I have begun
putting bees into the cellar from Nov. 4 to as late as Nov. 13.
This makes it necessary to have the bees hauled home by
about Nov. 1.

PUTTING BEES INTO THE CELLAR.

I like to get the bees into the cellar before the hives have
had a chance to contain any ice or damp combs from congeal-
ment of the bees' breath. It is also desirable to have the
outsides of the hives dry, so I do not like to have them go
into the cellar wet with snow or rain. Perhaps as good a
time as any is to get them in as soon as possible after a good
day for flying, commencing in the morning after a cold night;
preferably in clear weather. Often, however, I cannot have
everything just as I want it, and must take it as I can get it.
For 24 hours before taking in the bees, if not for several days
before, I open doors and windows of the cellars so as to give
them a good airing. One of the summer stands is put into
the cellar, having the back end raised 2 to 4 inches higher
than the front. This brings a hive, when placed upon this
stand, 3 or 4 inches from the ground at the front. If I had
abundance of room I would prefer to have them 6 inches
higher, as being more free from dampness and mold. The
cover is carefully lifted from a hive, so as not to disturb the
bees ; the hive is then carried into the cellar and placed on
the stand.

The hive not having been opened for some time before, the
bees have glued tight the quilt or cloth, so that on carrying
in, there is no change made on the top. The double hives
have the piece of lath taken away from the front, although
this is not always done, perhaps not generally done, till a
few days after they are in the cellar. After the hive is placed
on the stand in the cellar, a sheet of newspaper is placed over
it, then another hive, and so on till five hives are in the pile.
A hive-cover is placed on the hive at the top of the pile. The
reason for putting newspaper between each two hives in the

pile, is because, sometimes in the spring, when carrying out, a hive, when lifted from the pile, pulls the cloth or quilt with it from the next hive under it. The bee-glue of the quilt becomes attached to the hive over it, and the newspaper prevents this.

Thus in succession, additional piles are brought in, and each pile is independent of the others, so that if one happens to be jarred, only the hives in that pile are affected by it.

Often the bees get so warmed up by the middle of the forenoon, that they fly out when their hive is lifted to be carried into the cellar. In this case the hive is put back on its summer stand, and another colony, less wide-awake, is taken. But if the rousing up becomes general, operations must cease until the after-part of the day or the next morning. If for any reason, as the lateness of the season, or the fear of an approaching storm, it is thought best to carry in a hive whether the bees are willing or not, the entrance must be stopped. For this purpose,—as there is no danger of suffocation from stopping for a short time—I know of nothing better than a large rag or cloth which will easily cover the entire entrance. The rag must be dripping wet. In this condition it can be very quickly laid at the entrance, and being cold and wet the bees seem to be driven back by it, and when the rag is removed in the cellar, few if any bees come out. If dry, the bees would sting the rag, and upon its removal in the cellar a crowd of angry bees would follow it.

Sometimes the piles are ranged in a single row, clear around the cellar, the entrances facing toward the center and the backs 6 inches or more from the cellar wall. At other times they are placed in parallel rows across the cellar, two rows close together back to back. In every case a clear space is left in front of each hive so that I can easily approach it.

If, on carrying in, it should by any means happen that any hives are light, and there is fear that feeding may be necessary before they are carried out in the spring, such hives are piled separately where they are easily gotten at and toward spring a frame of honey is given each. As a general rule, however, I do not like tinkering at feeding in the cellar, and if a dozen hives had among them one that needed feeding, necessitating the opening and looking into the whole dozen, I thing I would rather run the risk of starving that one colony than to stir up the other eleven.

WARMING THE CELLAR.

"As yet, bee-keepers are not agreed as to the requisites for successful wintering, and I make no claim to a perfect knowledge of this-part of the business. I believe, however, that severe cold is had, and in this latitude, 42° north, I have known the mercury to reach 37° below zero. I now try to keep my cellars at not less than 45°, the thermometer being kept in the central part of the cellar. Sometimes the temperature gets down as low as 36° above, but not often and not for a long time. Oftener it stands at 50°. The heat is kept up by common small-cylinder stoves, having an inside diameter of about 8 inches between the fire-brick. There are two in the house-cellar and one in the shop-cellar. I burn hard coal in them, and by filling them up at morning and at night, this keeps a steady heat day and night, and there is not light enough to make any trouble. The expense is about $6 per stove for the winter. No matter how warm the weather, with very rare exceptions I keep the fires going at least lightly. The stove doors are always open."

In the spring, when there comes early a bright day with the mercury at 50° in the shade, and the temperature about the same in the cellar, it seems hardly necessary to keep a fire going; but I find by actual experience that I can keep the bees quieter with the fire. At least I feel pretty sure of it from what observations I have made. Probably that is due to the better ventilation caused by the fire.

VENTILATION OF THE CELLAR.

The ventilation of the cellar I consider a very important affair. No matter how well ventilated a hive may be, if the cellar in which it is placed contains nothing but foul air, how can there be good air in the hive? With good, pure air in a cellar, and an open entrance in each hive of 15½ by ½ inch, I do not feel much anxiety about ventilation. I am not sure but I should want a fire in a cellar for the sake of ventilation even if not needed for heat.

For the purpose of ventilation alone, the warmer the weather the more the fire in the cellar is needed. In zero weather, the air in the cellar, even where no fire is kept, is so much warmer and consequently lighter than the out-door air, that outer air, by its greater weight, forces itself into every crack and crevice of the cellar walls, displacing the lighter air of the cellar; and as fast as this fresh air becomes warmed, it is in its turn displaced by the outer cold air, and so a continuous change of air is kept up. Now suppose the air in the cellar stands at 40°, and the out-door air the same: there is nothing to change the air in the cellar. If the air in the cellar be now heated to 45° its increased lightness causes an influx of colder air from the outside, and the ventilation goes on as before. Of course there must be some limit to this, for when the temperature of the cellar goes above 60°, the bees show signs of uneasiness; although Mr. Ira Barber claims to have the temperature of his cellar sometimes as high as 90° without bad results.

The most difficult time to keep the bees quiet in the cellar, is when a warm spell comes in the fall soon after taking them in, or early in the spring. At such times I open up the cellar at dark. If very warm, all doors and windows are opened wide and by morning generally all are quiet. I leave all open as long as possible in the morning; sometimes till noon; when the bees begin to fly out all must be darkened.

I believe that it would be a good plan to have the cellar open
in such a way as would let in the air and keep out the light.

One morning, March 18, Emma went over to close the
shop-cellar; the doors having been open all night, and called
to me that there was as much as a colony of bees flying outside the cellar. The out-door air was 50°, in the cellar 53°,
and a very bright sun had been shining more than an hour.
Being broad daylight in the cellar, the bees were well
aroused, but there were not so very many outside—on the
wing they made a big show. I took a hive containing a
light colony and set it beside the door to catch the stragglers
which were kindly received by the colony.

SUB-VENTILATION OF CELLARS.

During warm spells it is more difficult to keep the bees
quiet in the shop-cellar than in the house-cellar. With
doors and windows of the former closed there is no pro-
vision for ventilation, except through the cracks, or the
opening of a trap-door overhead. In the house-cellar there
is a sub-ventilation pipe of 4-inch tile, 100 feet in length and
4 feet deep. This is, I think, quite inadequate for a cellar
having 31 by 33 feet for the outside measure of its walls, but
it is enough to make a very favorable difference. It seems
to me that the time may come when we shall understand
this matter of sub-ventilation so well that an abundance of
pure air will always be coming in at such a temperature that
there will be no need of artificial heat, and no need to pay
any attention to the bees from the time they are put in till
they are taken out. The entrance for air through the sub-
earth tube should be larger than the exit. Just how large
it should be for every hundred cubic feet of cellar room, how
long it should be and how deep, are matters that will perhaps be only fully learned by experiment.

MICE IN CELLARS.

Mice are troublesome denizens of cellars in winter. Even if they should be all cleaned out of the cellar before the bees are brought in, I am always sure to bring in some in the hives. I have fed them bountifully with bread and butter daintily covered with the various rat-poisons; have given arsenic mixed with sugar and flour; and have tried different traps, but still I find, every spring, holes gnawed by the mice in my nice, straight worker combs.

CLEANING OUT DEAD BEES.

Aside from attending to warming and ventilating my cellars, and unsuccessfully waging war against the mice, I think of no other attention given to the bees through the winter, except cleaning out the dead bees. For cleaning the dead bees out of those hives which have them—for some reason of which I am not yet sure, there are some hives which contain scarcely a dead bee—I have a very simple tool. It is a piece of round, ¼-inch or smaller iron rod, with one end hammered square for about two inches and bent at right angles, making something like a hook. With this hook I can reach into the hive under the frames and scrape out the dead bees.

I have a common kerosene hand-lamp with a sheet-iron chimney having a little mica window on one side—such as is used for heating water on lamps. This serves as a dark-lantern, making little light except in one direction. Holding the lamp in my left hand, I look in to see whether any live bees are in sight. Often I see the cluster near the front of the hive, oftener at the center or back part of the hive, the bees looking as if dead, so still are they; but in a few seconds some one will be seen to stir. In some hives nothing but dead bees can be seen, in some a lot of dead bees at the entrance with a few live bees crawling around among them,

and in others neither dead nor live bees can be seen,—
nothing but the bottom of the frames and a clean bottom-
board. In any case I scrape out all the dead bees that I can,
without disturbing the living. The upper hive in each pile
I clean standing; the other four I clean kneeling on an old
cushion to avoid mashing dead bees with my knees.

A whisk broom is used to brush off the entrance, and to
brush out the bees that are on the ground between the hives.
The bees on the ground are then swept up and carried out.
In the early part of the winter they need cleaning once a
month; later, once in two weeks.

I formerly thought that great care should be taken to avoid
any disturbance of bees in the cellar, and it seems to me now
that noise or jarring, or a light in the cellar, ought to be
injurious, but considerable observation makes me less afraid
of it. If the bees are very uneasy, a strong light upon them
will make them fly out, and at such times, as in a warm
spell in the spring, the cellar must be kept dark. I have,
however, had the sun shining brightly into the entrance of a
hive in the cellar without appearing to disturb the bees in
the least. This was on a morning after the cellar had
been thoroughly aired all night.

When it seems troublesome to keep the bees quiet during
a warm spell in early spring, the sun shining brightly, the
air still, and the mercury at 50° in the shade—just such a day
as would give the bees a splendid flight—there is a strong
temptation to bring them out; but if the soft maples are not
yet in bloom I believe they are better in the cellar; the next
day may bring cold, chilly winds, and many bees may be lost.

For instance: On March 19, 1886, the sun shone brightly
all day, and the weather had been warm for a few days
previous; but this day the mercury stood at 74° in the shade,
and even after dark at 66°, the air still—in every way a
delightful day. I thought it seemed too bad that the bees
could not enjoy it; but my wife reminded me of my own

teachings, that the maples were not out, and the bees not diseased. So I let the fires go down, and opened up the cellars at nights, the bees remaining in. Within 48 hours there came a blustering snow-storm, the mercury went down to 25°, I started the three fires, and the bees were not taken out of the cellar for a week later.

So long as the bees are not diseased, and can find no work to do abroad, their winter nap had better be continued.

EXTRACTING WAX.

Working, as I do, for comb honey, very little wax is produced. Bits of comb and wax are, however, constantly accumulating, and it is a nice thing to have this melted up, and out of the way. The most satisfactory thing I have found to melt up small amounts, is an old dripping-pan put in the oven of a cook-stove. The door of the oven is open, one corner of the pan projects out of the oven, this corner being torn open, the inside end of the pan is raised so that the wax as it melts may run out of the outside open corner, and a stone crock is placed under, to catch the dripping wax. In hot weather the pan and crock are put into a close box, out-doors, with a sash of glass over the box, and the sun does the work nicely. With reflectors properly arranged, the heat may be greatly increased, but I have never used anything but a common looking-glass; and that very seldom. Do not mash up brood-comb when you want to melt it.

OVERSTOCKING A LOCALITY.

To a bee-keeper who has more bees than he thinks advisable to keep in the home apiary, pasturage and overstocking are subjects of intense interest. The two subjects are intimately connected. They are subjects so elusive, so difficult to learn anything about very positively, that if I could well help myself, I think I should dismiss them altogether from contemplation. But like Banquo's ghost, they will not down.

I must decide, whether I will or not, how many colonies will
overstock the home field, unless I make the idiotic determi-
nation to keep all at home with the almost certain result of
obtaining no surplus. With what little light I have on the
matter, I do not care to have more than about 100 in one
apiary, although I do not know for certain that 125 or 150 in
a good year would fare much worse.

ARTIFICIAL PASTURAGE.

I have made some effort to increase the pasturage for my
bees. Of spider-plant I raised only a few plants. It seemed
too difficult to raise to make me care to experiment with it
on a larger scale. Possibly if I knew better how to manage
it, the difficulty might disappear. Or, on other soil it might
be less difficult to manage. The same might he said of the
other things I have tried. My soil is clay loam, and hilly,
although I live in a prairie State. I am at least a mile
distant from prairie soil. I have tried Alsike many times,
and never had a good stand but once; perhaps an acre then.
I had an acre of as fine figwort as one would care to see. It
died root and branch the second winter; even the young
plants that had come from seed the previous summer. It was
on the lowest ground I had, very rich, and much like prairie.

One year I raised half an acre of sun-flowers. Golden
honey-plant I never succeeded in getting to blossom. I
sowed perhaps 20 acres with melilot, and for the result I have
an acre or so of it growing. I doubt if I shall make any
further attempt to grow either of the plants I have mentioned,
except it be melilot. Wherever I have had a patch of it
started, it seems to hold its own from year to year. Possibly
also, I may try buckwheat, as some seasons I have had a
fair yield from that sown by others. Possibly, also, I may
set out some more basswood trees ; some twenty that were
set a few years ago produce a few blossoms now.

INDEX.

—o—

BOOKS FOR BEE-KEEPERS,

For Sale by Thos. G. Newman & Son, Chicago, Ills.

Bees and Honey, or Management of an Apiary for Pleasure and Profit, by THOMAS G. NEWMAN.—It is "fully up with the times," in all the various improvements and inventions in this rapidly-developing pursuit, and presents the apiarist with everything that can aid in the successful management of the honey bee, and at the same time produce the most honey in its best and most attractive condition. It embraces the following subjects: Ancient History of Bees and Honey—Locating an Apiary—Transferring—Feeding—Swarming—Dividing—Extracting—Queen Rearing—Introducing Queens—Italianizing—Bee Pasturage a Necessity—Quieting and Handling Bees—The Management of Bees and Honey at Fairs—Marketing Honey, etc. 220 profusely-illustrated pages. Price, in cloth binding, **$1.00**.

Apiary Register, by THOMAS G. NEWMAN.—This is a Record and Account Book for the Apiary, devoting two pages to each colony, ruled and printed, and is so arranged that a mere glance will give its complete history. Strongly bound in full leather. Price, for 50 colonies, **$1.00**; for 100 colonies, **$1.25**; for 200 colonies, **$1.50**.

Honey as Food and Medicine, by THOMAS G. NEWMAN.—It gives the various uses of Honey as Food; recipes for making Honey Cakes, Cookies, Puddings, Foam, Wines, etc. Also, Honey as Medicine, with many valuable recipes. It is intended for consumers, and should be liberally scattered to help in creating a demand for honey. Price, for either the **English** or **German** edition, 5 cents—one dozen, 40 cents—100 for **$2.50**—500 for **$10.00**—1,000 for **$15.00**. When 100 or more are ordered, we will print the bee-keeper's card (free of cost) on the cover.

Bee-Keepers' Convention Hand Book, by THOMAS G. NEWMAN. It contains a simple Manual of Parliamentary Law and Rules of Order, for the guidance of officers and members of Local Conventions—Model Constitution and By-Laws for a Local Society—Programme for a Convention, with Subjects for discussion—List of Premium for Fairs, etc. Bound in cloth, and suitable for the pocket. Price, 50 cents.

Why Eat Honey? by THOMAS G. NEWMAN.—This Leaflet is intended for distribution in the Bee-Keeper's own locality, in order to create a Local Market. Price, 50 cents per 100; 500 copies for **$2.25**; 1,000 copies for **$4.00**. When 200 or more are ordered at one time, we will print the honey-producer's name and address FREE, at the bottom. Less than 200 will have a blank where the name and address can be written.

Preparation of Honey for the Market, including the production and care of both Comb and Extracted Honey, and Instructions on the Exhibition of Bees and Honey at Fairs, etc., by THOMAS G. NEWMAN. This is a chapter from "Bees and Honey." Price, 10c.

Swarming, Dividing and Feeding Bees.—Hints to Beginners, by THOMAS G. NEWMAN. A chapter from "Bees and Honey." Price 5c.

Bee Pasturage a Necessity, by THOMAS G. NEWMAN—Progressive views on this important subject; suggesting what and how to plant.—A chapter from "Bees and Honey." 26 engravings. Price, 10c.

Bees in Winter, by THOMAS G. NEWMAN.—Describing Chaff-packing, Cellars and Bee Houses. A chapter from "Bees and Honey." Price 5c.

Bienen Kultur, by THOMAS G. NEWMAN.—In the **German.** Price, in paper covers, 40 cents, or $3 per doz.

The Hive I Use, by G. M. DOOLITTLE.—Price 5c.

Foul Brood, by A. R. KOHNKE.—Its origin and cure. Price, 25c.

BOOKS FOR BEE-KEEPERS,

EXCELSIOR HONEY-EXTRACTOR.

The $12.00 size.

Hundreds of unsolicited testimonials are received from those who are using these Extractors, highly commending them for ease of operating and general utility, and would respectfully refer all who are looking for the best Honey Extractor made, to any one possessing an Excelsior.

For 2 American frames, 13x13 inches..$8 00
For 2 Langstroth   "   10x18   "   . 8 00
For 3 Langstroth   "   10x18   "   ..10 00
For 4 Langstroth   "   10x18   "   ..14 00
For 2 frames of any size, 13x20   "   ..12 00
For 3   "   "   13x20   "   ..12 00
For 4   "   "   13x20   "   ..16 00

The eight and ten dollar sizes are made to accommodate those who desire a cheap but practical machine. The cans are smaller, the sides of the baskets are stationary and they have neither covers, strainers or metal standards.

Several improvements have been made over those of previous years, and points of excellence will be continually added to them as fast as discovered or suggested—keeping them fully up to the present advancing era of bee-keeping, and making the Excelsior an extractor only equaled by close imitation, and never excelled.

The Excelsior is made entirely of metal, and is consequently very light, strong and durable, with lugs at the bottom for firmly attaching to the floor, if desired.

The Comb Basket having vertical sides, insures the extracting power alike for top and bottom of frames. The sides of the basket in the $12.00, $14.00 and $16.00 Extractors being movable and Interchangeable, greatly facilitate the operation of dusting before and thoroughly cleaning after use, if desired. The basket can be taken from or replaced in the can in a moment, there being no rusty nuts to remove or screws to take out.

At the bottom of the can, and below the basket, is a cone or metal standard in $12.00, $14.00 and $16.00 Extractors, in the top of which revolves the bottom pivot of the basket, thereby giving room for sixty to one hundred pounds of honey without touching the basket or pivot below. The cans of the $8.00 and $10.00 sizes are shallow

THOMAS G. NEWMAN & SON,
923 & 925 West Madison Street, CHICAGO, ILLS.

www.ingramcontent.com/pod-product-compliance
Lightning Source LLC
Chambersburg PA
CBHW021937190326
41519CB00009B/1041